sextant

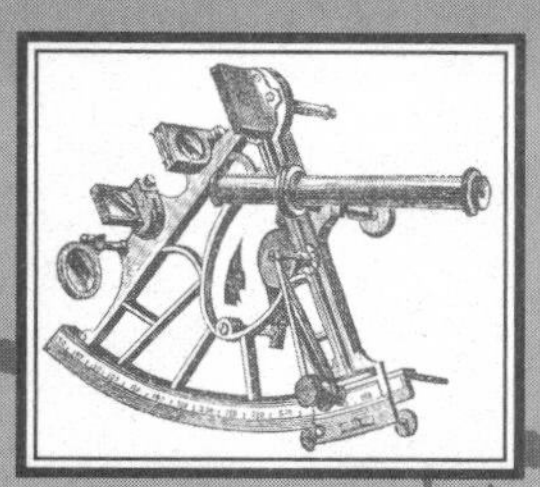

A Young Man's Daring Sea Voyage and the Men Who Mapped the World's Oceans

DAVID BARRIE

WILLIAM MORROW
An Imprint of HarperCollinsPublishers

Pages 319–320 serve as a continuation of the copyright page.

HarperCollins books may be purchased for educational, business, or sales promotional use. For information please e-mail the Special Markets Department at SPsales@harpercollins.com.

FIRST EDITION

Designed by Jamie Kerner

Library of Congress Cataloging-in-Publication Data has been applied for.

ISBN 978-0-06-227934-7

14 15 16 17 18 DIX/RRD 10 9 8 7 6 5 4 3 2 1

To the memory of my father, Alexander Ogilvy Barrie (1910–1969), who first showed me the stars, and of Colin McMullen (1907–1991), who taught me to steer by them

Contents

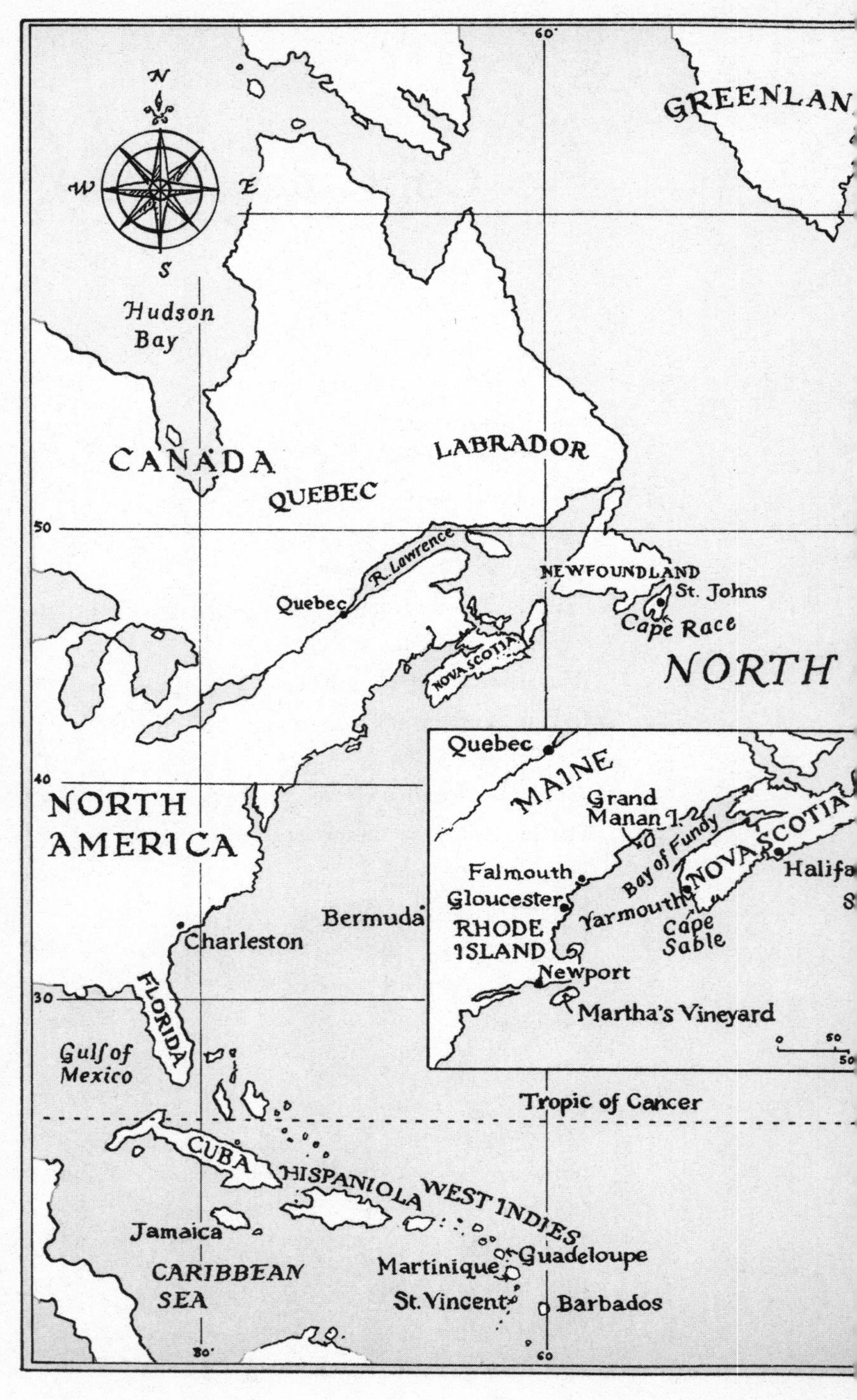
60°
N
W
E
S
GREENLAN
Hudson
Bay
CANADA
LABRADOR
QUEBEC
50
R. Lawrence
Quebec
NEWFOUNDLAND
St. Johns
Cape Race
NOVA SCOTIA
NORTH
Quebec
MAINE
Grand
Manan I.
Bay of Fundy
NOVA SCOTIA
Halifa
Falmouth
Gloucester
Yarmouth
RHODE
ISLAND
Cape
Sable
Newport
Martha's Vineyard
40
NORTH
AMERICA
Bermuda
Charleston
FLORIDA
30
Gulf of
Mexico
Tropic of Cancer
CUBA
HISPANIOLA
WEST INDIES
Jamaica
Guadeloupe
Martinique
CARIBBEAN
SEA
St. Vincent
Barbados
80°
60

ICELAND
Denmark Strait
Faroe Is.
Shetland Is.
Orkney Is.
Hebrides
NORTH SEA
DENMARK
IRELAND
ENGLAND
English Channel
Brest
FRANCE
Albi
ATLANTIC OCEAN
SPAIN
PORTUGAL
Lisbon
Gibraltar
Azores
Fayal
Sao Miguel
Madeira
Canary Is.
NORTH AFRICA
Cape Verde Is.
0 200 400 600 Km
200 400 M
20
60
50
40
30
20°
Tuskar Rock
St. George's Ch.
WALES
ENGLAND
Greenwich
The Smalls
Falmouth
Isle of Wight
Scilly Is.
English Channel
Channel Is.
0 270 Km
100 M
The NORTH ATLANTIC

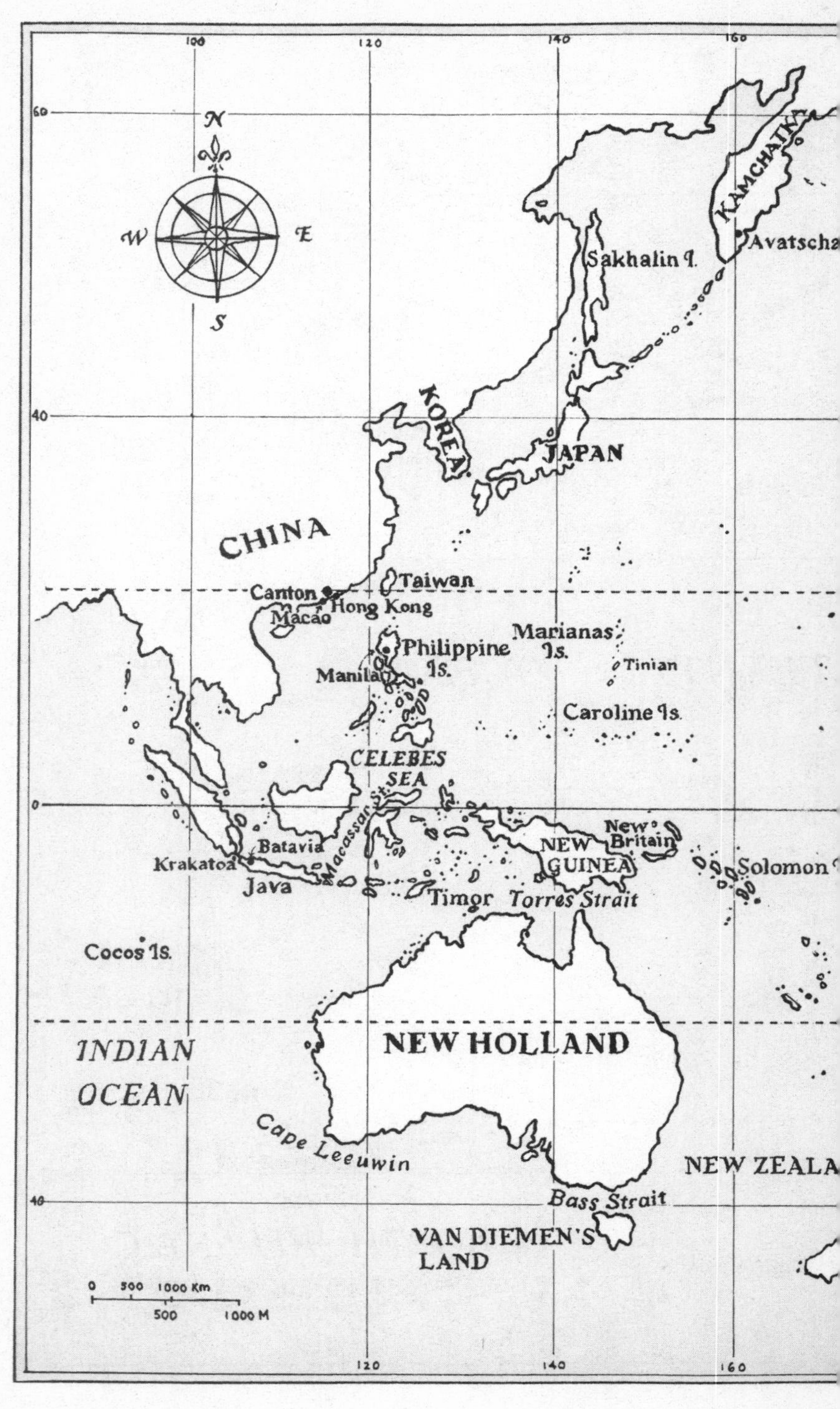
100
120
140
160
60
40
0
40
N
W
E
S
KAMCHATKA
Avatscha
Sakhalin I.
KOREA
JAPAN
CHINA
Taiwan
Canton
Hong Kong
Macao
Marianas Is.
Tinian
Philippine Is.
Manila
Caroline Is.
CELEBES SEA
Macassar St.
Batavia
Krakatoa
Java
NEW GUINEA
New Britain
Solomon
Timor
Torres Strait
Cocos Is.
NEW HOLLAND
INDIAN OCEAN
Cape Leeuwin
NEW ZEALA
Bass Strait
VAN DIEMEN'S LAND
0 500 1000 Km
500 1000 M
120
140
160

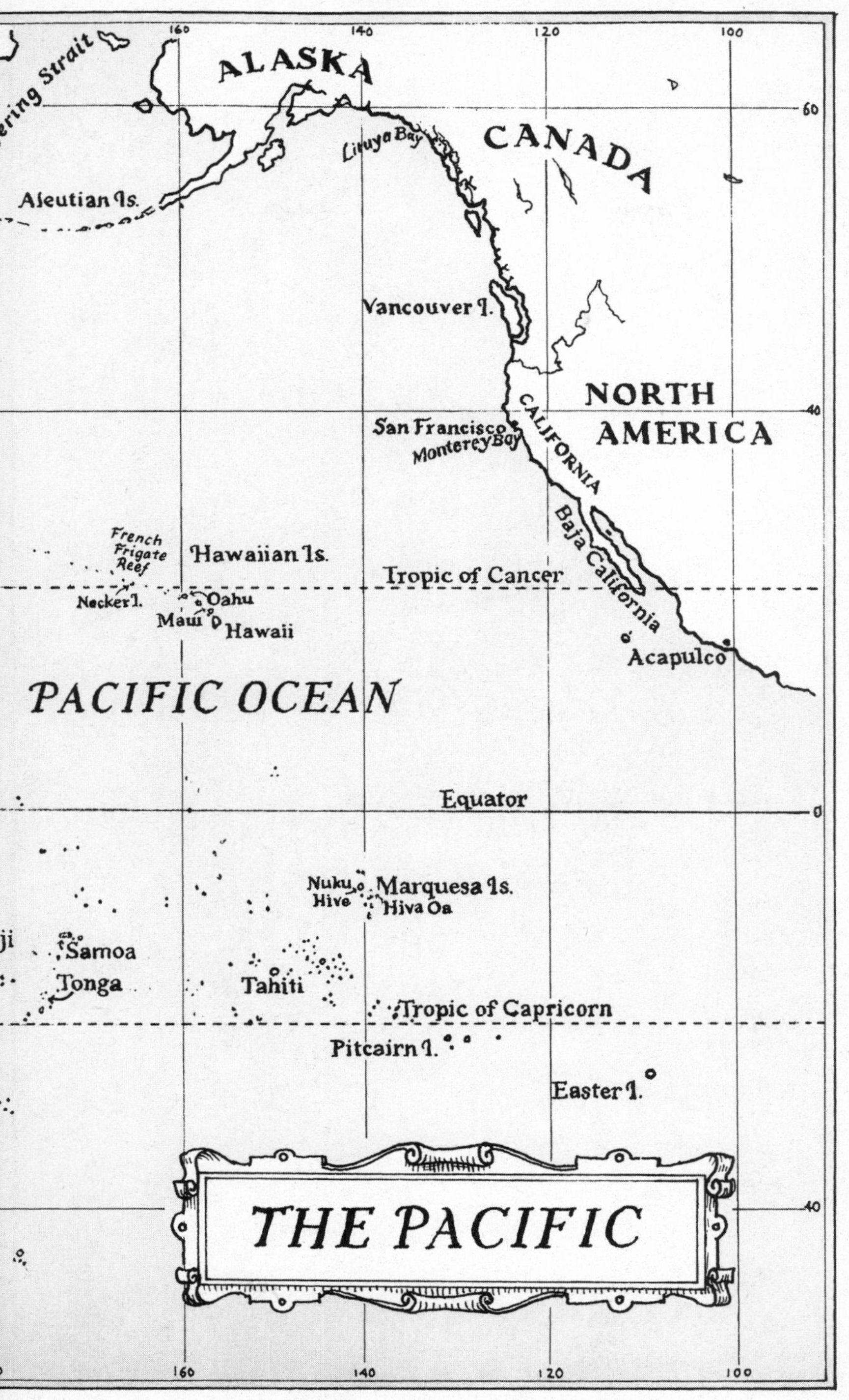

THE PACIFIC
ALASKA
Bering Strait
CANADA
Lituya Bay
Aleutian Is.
Vancouver I.
NORTH AMERICA
CALIFORNIA
San Francisco
Monterey Bay
Baja California
French Frigate Reef
Hawaiian Is.
Tropic of Cancer
Necker I.
Oahu
Maui
Hawaii
Acapulco
PACIFIC OCEAN
Equator
Nuku Hive
Marquesa Is.
Hiva Oa
Samoa
Tonga
Tahiti
Tropic of Capricorn
Pitcairn I.
Easter I.
160
140
120
100
60
40
0
40

Introduction

Crossing an ocean under sail today is not an especially risky undertaking. Accurate offshore navigation—for so long an impossible dream—has now been reduced to the press of a button, and most modern yachts are strong enough to survive all but the most extreme weather. Even if errors, accidents, or hurricanes should put a boat in danger, radio communications give the crew a good chance of being rescued. Few sailors now lose their lives on the open ocean: crowded inshore waters where the risk of collision is high are far more hazardous.

But it was not always so. When a young man called Álvaro de Mendaña set sail from Peru in November 1567 to cross the Pacific with two small ships, accompanied by 150 sailors and soldiers and four Franciscan friars, he faced difficulties so great that his chances of survival, let alone achieving his objectives, were slim.[1]

Mendaña's orders from his uncle, the Spanish viceroy, were to convert any "infidels" he encountered to Christianity, but the expedition was certainly not motivated entirely by religious zeal. According to Inca legend, great riches lay on islands somewhere to the west. Were these islands perhaps outliers of the great southern continent that was believed to lie hidden somewhere in the unexplored South Seas? Mendaña, who was twenty-five, hoped to find the answer, to set up a new Spanish colony, to make his fortune and win glory. However, any optimism he may have felt as the coast of Peru dipped below the horizon would have been misplaced. Although Magellan had managed to cross the Pacific from east to west in 1520–21, he had been killed in fighting with local people after reaching the Philippines, and only four out of the forty-four men who sailed with him aboard his small flagship had returned safely to Spain.[2] This first, epic circumnavigation was counted as a brilliant success, but other expeditions ended in oblivion.

The challenges Mendaña faced were many. Not only was it impossible to carry sufficient fresh food and water for a voyage that might well last several months, but sailing ships were also vulnerable to the stress of weather, and the discipline of their rough and uneducated crews could never be relied on. First encounters with native peoples were fraught with danger, even if both sides were keen to avoid conflict, not least because cultural and linguistic differences made communication so difficult. If the Europeans brought with them infectious diseases that were to devastate native populations, tropical diseases also posed a serious threat to the visitors. To venture into the unexplored wastes of the Pacific was therefore to risk shipwreck, mutiny, warfare, disease, thirst, hunger, and, most insidious of all, malnutrition.

After a passage of eighty days, Mendaña's two ships at last reached the "Western Islands" in February 1568. Thinking at first that they had indeed found the legendary southern continent, Mendaña and his men explored the high, jungle-clad island on which they first landed and soon realized their mistake. They named it Santa Isabel, because they had sailed from Peru on that saint's feast day, and went on to visit the neighboring islands, which they called Guadalcanal, Malaita, and San Cristóbal. Though a chief had greeted the Spanish visitors warmly on their first arrival, the natives could not satisfy their pressing demands for food; Mendaña had difficulty controlling his men—and blood, mostly native, soon flowed.

In August a disappointed Mendaña set sail from San Cristóbal. Having barely survived a hurricane, Mendaña and his officers had no idea where they were, how far they had travelled, or when they might again reach land. Their few navigational tools would have included astrolabes and quadrants for determining latitude, magnetic compasses to steer by, hourglasses for measuring short intervals of time, and lead-lines for sounding the depth in shallow water. But they had no proper charts and—crucially—no reliable means of judging how much progress they had made either to the east or to the west: only by estimating the ship's speed through the water could the pilots assess how far they had travelled. This was a deeply unreliable method.

The agonizingly long return journey took Mendaña in a wide circuit across the North Pacific to reach the coast of Baja California in December 1568. He and his crew were reduced to a daily allowance of six ounces of rotten biscuit and half a pint of stinking water. Scurvy swelled their gums until they covered their teeth, they were racked by fever, and many went blind. Every day they had to throw overboard another corpse. It was not until the following September that Mendaña finally reached Peru. He had found no riches, no continent, had made not a single convert, and had failed to establish a colony, but his extraordinary voyage was to become a legend. Though he had been obliged to mortgage his property to get his ship repaired in Mexico, rumors spread that he had come home laden with gold and silver.

The islands he had discovered were soon known by the name of the fabulously rich king of the Old Testament: Solomon.[3] The longitude that he assigned to the Solomon Islands was so wildly inaccurate that subsequent explorers repeatedly failed to find them and eventually began to doubt their existence.[4] It was to be two hundred years before any European set foot on the Solomons again.

Mendaña himself failed to find the islands he had discovered when he mounted another, completely disastrous transpacific expedition in 1595, accompanied by Pedro Fernández de Quirós as chief pilot. He died in the Santa Cruz Islands—pathetically close to his goal—and Quirós eventually brought the disappointed survivors home to Peru via Manila after "incredible hardships and troubles."[5] Later generations of mariners and cartographers, deprived of detailed information about these voyages by the secretive Spanish authorities, struggled to make sense of Mendaña's claims, and the Solomon Islands shifted giddily about the Pacific, varying in longitude by thousands of miles and even in latitude from 7 degrees to 19 degrees South. In 1768, within the space of a few months, two European mariners—Carteret and Bougainville—passed among the Solomons again, but without even realizing that they were following in Mendaña's wake.[6] They were soon followed by a French trader, Jean de Surville (died 1770), who visited the islands in 1769. Having closely investigated the accounts of these voyagers, and compared them to the descriptions that Mendaña had given, Jean-Nicolas Buache de Neuville (1741–1825)[7] understood that the Solomon Islands had at last been rediscovered, though his arguments were not immediately accepted. His fellow countryman La Pérouse was to lose his life in trying to confirm his theory. Rear Admiral Joseph-Antoine Bruny d'Entrecasteaux finally settled the matter when searching for La Pérouse in the 1790s. He recognized many of the islands that Mendaña had described and decently restored to them the Spanish names that had been bestowed on them so long before.

The finding of the Solomon Islands, their subsequent "disappearance," and their eventual rediscovery perfectly illustrate the difficulties that confronted transoceanic navigators of the early modern age. It would be easy to multiply examples of this kind, which reveal the intimate, reciprocal dependence of navigation and hydrography—a recurrent theme of this book. The point is a simple one, but easily overlooked. To find the way safely, a mariner needs a chart that accurately records the positions of all that is navigationally significant—from the outlines of the major landmasses to the precise locations of tiny, uninhabited shoals on which a ship could founder. To make such charts, however, the hydrographer must first know the exact positions of everything that is to appear on them. Hydrography serves navigation, but only if nourished first by the fruits of navigation.

Two hundred and fifty years ago it was not just the location of the Solomon Islands that lay in doubt. Though it is hard for us to imagine such a state of affairs, the shapes of whole continents then remained largely unknown, and accurate charts—even of European waters—did not exist. The main reason for this state of ignorance was the imperfection of the art of celestial navigation and in particular the impossibility of determining longitude with any precision on board ship. In 1714 an act of Parliament was passed in Great Britain designed to encourage the development of a practical shipboard solution to this age-old problem. It was not the first such prize but it turned out to be the last. Within fifty years, and in the space of a single decade, two radically different solutions emerged, one mechanical and the other astronomical (see Chapter 6). The long-running and often ill-informed tussle between the advocates of these two methods has obscured the fact that *both* depended on a newly developed observational instrument: the sextant.[8] Though its praises have seldom been sung, the sextant was to play a crucial part in shaping the modern world—both literally and figuratively.

THE SEXTANT, LIKE the anchor, is a familiar symbol of the maritime world, but to most people—including many sailors—its purpose is a mystery. One strand of my task is to sketch the developments in astronomy, mathematics, and instrument making that first permitted navigators to fix their position with its help. But I also wish to bring the sextant to life by examining some of the astonishing feats of the explorers who put this ingenious instrument to such good use in making the first accurate charts of the world's oceans. The work of the pioneering marine surveyors of the late eighteenth and early nineteenth centuries—some of whom are almost forgotten—is another key strand. Because it is such a wide subject, I have focused on those who worked in the Pacific, which was then the subject of greatest interest; the examples I have chosen illustrate some of their most remarkable achievements* as well as the many challenges they faced.[9] I have also squeezed in the stories of three exceptional small-boat voyages, each of which depended crucially on skillful celestial navigation: Captain Bligh's journey from Tonga to Indonesia after the *Bounty* mutiny, Joshua Slocum's circumnavigation of the world in his yacht *Spray*, and Sir Ernest Shackleton's remarkable rescue mission crossing the Southern Ocean in the *James Caird*, piloted by Frank Worsley.

To speak of the "discovery" by European navigators of lands that had long been inhabited by other peoples is obviously absurd, if not insulting, but since the focus of this book is a European invention, Europeans unavoidably take center stage.

---

* The sextant was also a crucial aid to land-based explorers and indeed aviators; their stories, however, lie beyond the scope of this book.

By way of contrast, I have mentioned briefly the extraordinary skills of the Polynesian navigators, who found their way across the wide expanses of the Pacific using neither instruments nor charts long before the arrival of Western explorers. Their achievements deserve to be better known, but they have been well described by others,[10] and this is not the place in which to discuss them more fully.

This is not a "how to" guide to celestial navigation, but I hope I have given enough information to enable the reader to grasp its basic principles. I have also tried to give some sense of what it *feels* like to navigate across an ocean in the old-fashioned way, with sextant and chronometer. Many of the great explorers who wrote about their experiences did so for fellow professionals who needed no explanations, while those who addressed the general public must often have supposed that descriptions of celestial navigation would make dull reading. Anecdotes from Slocum and Worsley have helped me to fill this gap, but I have also drawn on my own—far more modest—experiences, including those recorded in a journal I kept when sailing across the Atlantic as a teenager forty years ago.

For two hundred years mastery of the sextant was a vital qualification for every oceangoing navigator. Hundreds of thousands of young men (women seldom had the opportunity) worked hard to learn the theory and practice of celestial navigation, and experts wrote manuals that sold in large numbers to cater to their needs. But the use of the sextant is now an endangered skill that is most commonly learned only to provide a safety net should the now-ubiquitous Global Positioning System (GPS) fail.[11] Very few practice taking sights at sea as a matter of routine, and most sailors rely almost entirely on electronic navigation aids. The sextant, if not yet forgotten, has been relegated to a very occasional understudy role. Almost without notice, the golden age of celestial navigation has drawn to a close.

If this book has an elegiac tone, it is not merely an exercise in nostalgia. I hope and believe that the sextant has a useful future—that it is not destined to join the many outmoded scientific instruments preserved only in museums. It would, of course, be more than a little eccentric to dismiss the convenience (and reassurance) of electronic guidance systems, but reading off numbers from a digital display is a very thin, prosaic experience compared with the practice of celestial navigation. GPS banishes the need to pay attention to our surroundings, and distances us from the natural world; although it tells us precisely where we are, we learn nothing else from it. Indeed, unthinking reliance on GPS weakens our capacity to find our way using our senses. By contrast, the practice of celestial navigation extends our skills and deepens our relationship with the universe around us.

What could be more wonderful than to join the long line of those who have found their way across the seas by the light of the sun, moon, and stars? Just as interest in classic boats, built in traditional ways and shaped only by the demands of beauty and seaworthiness, has undergone a revival, so the joys of navigating with a sextant are now ripe for rediscovery.

sextant

Chapter 1

Setting Sail

Sextant: I was nine years old when I first heard that magical word. It was 1963 and I had gone with my family to see *Mutiny on the Bounty*, starring Trevor Howard as the notorious Captain Bligh, whom he played as a choleric middle-aged martinet, and Marlon Brando as his infuriatingly condescending, toffee-nosed first officer, Fletcher Christian.* A luscious, big-budget movie, shot in the South Pacific around Tahiti, it ends with the burning of the *Bounty* by some of the mutineers after their arrival at the remote (and then incorrectly charted) Pitcairn Island. Christian, the leader of the mutiny, tries in vain to save the ship and, before abandoning it, calls out over the roar of the flames to his friend:

Fletcher: "Have you got the sextant, Ned?"

Ned [unable to hear]: "What?"

Fletcher [shouting desperately]: "Have you got the sextant?"

Ned: "No!"

[Fletcher dashes for the companionway that leads to the captain's cabin, below the burning decks]

Ned [yelling in alarm]: "You can't go now—it's too late, Fletcher!"

Fletcher [rushing below regardless]: "We'll never leave here without it!"

---

* Actually, Christian was master's mate.

Christian dives into the blazing cabin and is horribly burned trying—in vain—to recover the precious instrument, later dying on the shore as the ship goes down in a shower of steam and sparks.

MY FATHER LOVED astronomy and, as a civil engineer, he had been trained in surveying and mapmaking. It was he who first showed me the night sky when I was a very small boy, standing in our Hampshire, England, garden on many cold, clear winter nights beneath the dark Scots pines. He taught me to recognize the flattened W of Cassiopeia, the great torso of Orion, and Ursa Major (the "Big Dipper") with its twin pointers—Dubhe and Merak—that lead the eye to the North Star: Polaris. The Milky Way, I learned, was a galaxy composed of billions of stars to which our sun and solar system belonged as just one very small element.

As we left the cinema I asked my father what a sextant was, and why it mattered so much. I do not remember exactly what he said, but I gathered that it was a device for fixing your position anywhere in the world, on land or sea, by reference to the sun and stars—and that it was a vital tool for navigators sailing out of sight of land. Coupled with the terrifying image of Fletcher Christian diving into the inferno, his words caught my imagination: the thought of being marooned forever on a small, remote island, unable ever to find the way home, was haunting. How could so much depend on one small instrument? And how could the unimaginably distant sun and stars help a sailor find his way across a vast ocean?

This was the beginning of my fascination with the art of navigation. I lived in a town on the south coast of England where sailing was a part of everyday life, and I first went out in an old-fashioned clinker-built dinghy with my parents when I was not much more than a toddler. I still remember dozing off on a sail bag, tucked up under the half deck, listening to the slap of the water on the bows, hypnotized by the gentle, broken rhythm of the waves. Later I sailed a dinghy of my own and crewed racing yachts, but I never much liked competitive sailing. What I loved was pilotage—the business of reading a chart, plotting a course, making allowances for compass variation and the effects of tidal streams, and all the other tricks of the coastal navigator's trade.

Charts fascinated me. Those published by the British Admiralty were then still printed from engraved plates, and their appearance had not changed much since the nineteenth century. They had a solemn gravity, reflecting as they did the accumulated data of generations of dedicated marine surveyors. The traditional saying—"Trust in God and the Admiralty chart"—was a measure of their exalted reputation. Unlike their metric successors, the old charts were soberly black-and-white. Prominent features on dry land that might be useful to the navigator—like church steeples or mountains—were shown, and detailed views of the coast were often included in the margins to aid recognition of important landmarks or hazards: the old surveyors were all trained as draftsmen. Wrecks were marked with a variety of warning symbols depending on how much water covered them; those that broke the surface even at high water were marked with a grim little ship, slipping stern-first beneath the waves. The nature of the "ground" (that is, the seabed) was indicated in a simple code—"m" for mud, "sh" for shingle, "s" for sand, "rk" for rock, "co" for coral, and so on. Charts vary enormously in scope: the large-scale ones of harbors might cover an area of only a few square miles, while others cover entire oceans. The smaller-scale ones are framed by a scale of degrees and minutes of latitude (north–south) and longitude (east–west), and the surface is carved up by lines marking the principal parallels and meridians—an abstract system of coordinates first conceived by Eratosthenes (*c.*276–194 BCE) and then refined by Hipparchus (*c.*190–120 BCE). Compass "roses" help the navigator to lay off courses from one point to another and show the local magnetic variation—the difference between true north and magnetic north.

From my father I learned something about surveying and the use of trigonometry—the mathematical technique for deducing the size of the unknown angles and sides of a triangle from measurements of those that are known. On our walks in the New Forest we sometimes came across the concrete triangulation pillars on which the British Ordnance Survey maps were based. Each pillar formed the corner of a triangle from which the other two corners were visible. Starting from a very accurately measured baseline, a network of such triangles extended across the whole country. By measuring the angles between the pillars using a theodolite, surveyors could determine the relative positions of each pillar with great accuracy, thereby providing the mapmakers with an array of fixed points on which to build. In those days this system was still the key to land-based cartography.

Marine charts were liberally sprinkled with "soundings"—numbers representing the depth of water in old-fashioned fathoms (1 fathom to 6 feet), which crowded in even greater profusion around hazardous patches of sea. Particularly sinister were the places in the mid-ocean depths where a tight cluster indicated an isolated shoal—perhaps the tip of a "sea mount" that did not quite break the surface. The Chaucer Bank, some 250 miles north of the Azores in the middle of the North Atlantic, is an example. On Admiralty chart no. 4009 (North Atlantic Ocean—Northern Portion, published in 1970) it rose up to a "reported" minimum depth of 13 fathoms from waters that slide down rapidly to 1,000 fathoms or more. In heavy weather, seas would break on such a shoal—an alarming sight so far from land, and a potential hazard, too. Before the advent of the electronic echo sounder in the 1920s, all these soundings would have been taken with lead-lines—nothing more than a lump of lead on the end of a long, calibrated rope or wire. Triangulation could have been used to fix the positions of soundings along the coast, but what about those offshore, far out of sight of land? Of the vital part the sextant had played in hydrography—the mapping of the seas—I had as yet no idea.

As a teenager I sailed to Normandy and Brittany and around the west coasts of Ireland and Scotland. These excursions offered plenty of navigational challenges—the English Channel with its strong tidal currents and heavy shipping traffic is a dangerous stretch of water and the many rocky shoals of Brittany, Ireland, and Scotland demand respect—but they did not call for the use of a sextant. Instead we relied on dead reckoning (DR—using the distance travelled and the course followed to estimate your position) corrected by radio direction-finding (RDF—fixing the boat's position by taking compass bearings of radio beacons). If sailing at night, compass bearings of lighthouses were helpful, too. While these methods worked well enough for short coastal passages, I wanted to know more: I was determined that one day I would learn how to navigate the open ocean by the sun and stars. I had not yet even seen a sextant, but the mysteries of celestial navigation already had me under their spell.

JUST TEN YEARS after seeing *Mutiny on the Bounty*, I got my first chance to handle a sextant when a family friend invited me to help him sail across the North Atlantic in his thirty-five-foot sloop, *Saecwen*.* Colin McMullen was a retired Royal Navy captain, and like many naval officers he was easygoing, relaxed,

* She was a "Saxon" class boat, designed by Alan Buchanan and built by Priors of Burnham-on-Sea, in 1961. Her name means "sea queen" in Anglo-Saxon.

and charming—useful if not essential qualities when sharing cramped accommodation for any length of time. Colin loved nothing better than an impromptu party. On the slightest pretext he would get out his accordion and start a "sing-song," and if he was in particularly high spirits he might even put on a false beard and impersonate an ancient mariner with a strong west-country accent.

Colin was also fond of practical jokes, one of which almost cost him his life. As a young midshipman on board a small yacht being towed by a much larger vessel, he decided it would be amusing to climb along the tow rope and appear—as if by magic—on the deck of the mother ship. This meant scrambling along a heavy hawser, the middle of which frequently dipped beneath the surface of the sea. Colin was barely able to hold his breath long enough and nearly lost his grip as the cold, fast-moving water tugged at his submerged body. He was carpeted for this crazy escapade, but in the Royal Navy of the 1920s there was room for colorful characters, and it did his career no harm.

Colin had been messing about in boats since his childhood days at Waterville in County Kerry, Ireland, during World War I. When he was posted to Malta in the 1930s he was given the enviable task of delivering the commander-in-chief's official yacht to Venice, and I remember him talking rapturously about the summer days he spent along the Croatian coast aboard this large and elegant vessel. Most of his sailing, however, had been on a much more modest scale—notably in a small yacht called *Fidget*, which he shared for a time with a group of fellow naval officers.

Colin bought *Saecwen* after retiring from the navy, and I first crewed for him when the two of us sailed her along the south coast of England from Dartmouth to her home port, Lymington, in early January 1972. It was an overnight trip and the weather was clear, cold, and windless. As we motored slowly across the wide expanse of Lyme Bay I watched the "loom" of French lighthouses, one of which—on the notorious Roches Douvres reef, off the coast of Brittany—was nearly eighty miles away, far beyond the range at which it would normally be visible. The distant pencil beam of light rose briefly from below the horizon, sweeping up and over like the headlights of a car making a sharp turn on the far side of a hill.*

In the middle of my watch I heard the hatch slide back, and there was Colin—who should have been asleep—with two cups of hot cocoa. While I steered we sat together looking up at the night sky, our breath smoking in the cold. It was then that we first talked about celestial navigation. Colin pointed out the stars to me and recalled his days as a young naval cadet, just after World War I, when learning to handle a sextant and plot a line of position had been nothing but a chore. Now he was planning a transatlantic cruise in *Saecwen* and was looking forward to brushing up his old skills. Nothing was said at the time, but later that year Colin asked if I might be free for six weeks or so the following summer; the trip to America was going ahead and he was looking for crew on the return voyage to England. As a university student with time to spare I eagerly accepted the invitation: here was a chance not only to cross an ocean under sail but also to learn the art of celestial navigation from a professional whom I admired. But transatlantic passages in small boats were not yet the fairly routine events they have since become. Looking back I am amazed that my mother, who had been widowed not long before, raised no objections. She must have felt the risks were worth taking.

---

* Each lighthouse has its own "characteristic" by which it can be identified—in this case a single flash every five seconds.

ON A STICKY evening in early July 1973 I arrived at Falmouth, a small town on the coast just north of Portland, Maine. It was my first visit to the United States, and I had travelled up from New York on a Greyhound bus. The license plates on the cars announced that I was in "Vacationland," the temperature was in the 90s, and the humidity was only slightly less than in Manhattan, though the still air was refreshingly clean. From the bus terminal I took a cab to the Portland Yacht Club, and as I walked down the jetty I caught sight of *Saecwen* lying at a mooring only fifty yards away. Rocky islands covered with hemlock and spruce lay farther offshore. Colin was watching out for me and pulled across at once in the rowing dinghy to pick me up. It was strange to step aboard *Saecwen* again in such different surroundings, but as the deck gently rocked beneath my feet I felt almost as if I had come home. I slung my bag into the starboard quarter berth where I was to sleep for the coming weeks, absorbed the familiar smells, and came up on deck where Colin handed me a can of beer—very welcome in that heat. He was not alone on board *Saecwen*. There were two other crew members at this stage—Colin's sister, Louise de Mowbray, and his cousin, Alexa Du Vivier, who was just seventeen. We talked about *Saecwen*'s voyage out across the Atlantic—it had been tough going, with several gales, and one member of the crew had suffered so badly from seasickness that he had been forced to leave the boat in the Azores. Transatlantic passages by small yachts were then sufficiently rare that *Saecwen*'s British ensign had caused a lot of excitement when she arrived in Gloucester, Massachusetts. The news media had soon discovered Colin, who, playing the role of old British sea dog to perfection, had been interviewed on TV and radio. He was delighted by all the attention.

We set sail the next morning. Wherever *Saecwen* dropped anchor, as we cruised north and east through the rocky, wooded islands that sprinkle the coast of Maine, complete strangers appeared offering food, showers, and lifts to the shops—and we did not hesitate to accept. I started to read Samuel Eliot Morison's classic account of the early European voyages of discovery to the Americas, which I found on board. I learned—to my surprise—that it was French rather than British mariners who had first properly charted the seaboard along which we were now sailing. The northeast coast of America is notorious for its cold fogs, but luckily we encountered little until we set out across the Bay of Fundy, from Grand Manan Island, past Yarmouth toward Cape Sable at the southern end of Nova Scotia. Fog is strangely disorienting, and at times it was so thick that we could hardly see the bows of the boat from the cockpit. Colin was busy plotting our position by RDF as we rounded Cape Sable by night, when we had a bizarre encounter with a ferry whose approach first became apparent when we heard the distant beat of pop music. It grew steadily louder until at last the thump-thump of the ship's propellers also became audible. By now we were really worried, but there was little we could do to reduce the risk of a collision apart from tooting feebly on our small foghorn, all too well aware that we had little or no chance of being heard. Suddenly the blurry outline of a big ship, brilliantly illuminated, emerged from the fog, and the air was filled, bizarrely, with Bill Haley's "Rock Around the Clock."

The ferry shot past us at a distance of no more than a quarter of a mile, disappearing very quickly, while the music gradually faded. We were carrying a radar reflector and ought to have been plainly visible on the ship's radar, but we had the uncomfortable feeling that no one had seen us—and we knew that if we had been run down, the slight bump would probably have passed unnoticed. There was an inflatable life raft lashed to the foredeck, but even if we could have launched it in time, in those cold waters rescue would have had to come quickly. On the open sea, collision is the biggest risk faced by a well-managed yacht, as we were dramatically reminded a few weeks later, far out in the Atlantic.

OUR LAST PORT of call was the magnificent and historic natural harbor of Halifax, Nova Scotia, where we spent a week busily preparing for the crossing. Louise now flew home, leaving just Colin, Alexa, and me aboard *Saecwen*.

Halifax served as the Royal Navy's main base in America during the Seven Years' War, a pan-European conflict that formally broke out in 1756 (though hostilities between the British and French and their native allies had already begun in America) and was fought in many different parts of the world. It was here that James Cook—who was to become one of the greatest European explorers—began to learn the science of surveying from Samuel Holland, a military engineer serving with General James Wolfe.[1]

In 1759 Cook played a key part in the daring survey work in open boats on the narrow and dangerous "Traverses" of the St. Lawrence River below Quebec City. The safe channels having been marked, the British fleet was able to pass the Traverses without a single loss, thereby permitting Wolfe to land his forces upstream of Quebec. The French governor angrily commented: "The enemy have passed 60 ships of war where we dare not risk a vessel of 100 tons by night or day."[2] Wolfe and his French opposite number, Louis-Joseph de Montcalm, both lost their lives in the famous battle on the Plains of Abraham that followed, but Quebec fell to the British, and the final expulsion of the French from North America soon followed.

Cook, who had not yet received an officer's commission, continued surveying throughout his time on the American coast, and in 1761 he was given a bonus of £50 "in consideration of his indefatigable Industry in making himself Master of the Pilotage of the River St Lawrence &c"—a most unusual distinction.[3] The following year, while assisting in the recapture of the port of St. John's, Newfoundland, Cook worked with another remarkable military engineer, Joseph DesBarres, and carried out important surveys that brought him to the attention of the Admiralty in London.[4] He was starting to make his name.

COLIN'S ORIGINAL PLAN had been to carry on to St. John's, Newfoundland, calling on the way at the small islands of St. Pierre and Miquelon—the last remaining French outposts in North America, which Cook surveyed in 1763[5] just before they were returned to France at the end of the Seven Years' War. However, the icebergs emerging from the Arctic had drifted much farther south than usual in the summer of 1973, and we decided that it was wiser not to go any farther north. Dodging icebergs in a wooden boat is risky. The most dangerous kind are the "growlers"—small pieces of ice but still weighing many tons, completely awash and therefore almost invisible above the surface—and bumping into one of them might have brought our voyage to a quick and fatal conclusion.

The last couple of days in Halifax were filled with lists. There were loads of provisions to buy, including a whole chicken in a tin to be saved for a special occasion. Everything had to be carefully stowed in one of *Saecwen*'s many lockers, and a record kept of its location. Having removed their paper labels—which might well get washed off—we marked the tinned goods with a waterproof felt pen. Fresh vegetables and fruit went into cargo nets hanging from the low cabin roof. We checked the rigging for signs of wear, and I was hoisted in a bosun's chair to make sure that all was well at the masthead. Looking down from that height, *Saecwen* seemed very small indeed. Finally we did our laundry, filled up with diesel fuel, paid the harbor dues, and said farewell to our Canadian friends. Early the next day we topped up the water tank, cast off, and motored down the harbor.

> *The taking of Departure* [*wrote Joseph Conrad*], *if not the last sight of the land, is, perhaps the last professional recognition of the land on the part of a sailor. . . . It is not the ship that takes her departure; the seaman takes his Departure by means of cross-bearings which fix the place of the first tiny pencil-cross on the white expanse of the track chart, where the ship's position at noon shall be marked by just such another tiny pencil-cross for every day of her passage.*[6]

We took our Departure from a whistle buoy just off the harbor entrance. Thousands of miles of ocean and weeks of sailing lay ahead of us. We had no way of telling what weather we might face and would not be able to receive forecasts. We could only keep an eye on the barometer and hope for the best. I felt like an actor stepping on to the stage at the start of a big performance as I hauled up the sails. We hardened the sheets and Colin cut the engine. Apart from the sound of the wind and waves, all was quiet. *Saecwen* heeled to the southeasterly breeze and began to dip her bows into the Atlantic swells. Cold spray rattled over my oilskins. Here we go, I said to myself.

Chapter 2

First Sight

Days 1–2: Didn't sleep much. Up at 0730 for breakfast very conscious it would be the last meal on dry land for a while. Took shower in the club house. Our Canadian friends came to see us off and we set sail at 0930 for England. There was little if any wind and a thick cold fog soon rolled in. At 1115 we heard the outer whistle buoy close by. The wind picked up from SE about 4–5 but with a confused sea. Alexa felt sick and went below and I started feeling queasy, too.

Colin and I had tinned beef stew and potatoes for supper. Now seem to be getting my sea legs back but Alexa still curled up in the fo'c's'l out of action neither eating nor drinking.

I took the watch from midnight until 0400. The far end of my sleeping bag was soaking wet when I turned in. Up again at 0800. Dull, grey morning, but no more fog.

Bumpy seas, as we plugged on to the south of Sable Island under reefed main. At 1200 things brightened up and the sun came out. Wind eased to force 3. Put up genoa [the largest foresail] and got a bit more speed. Had a nap before supper, then took watch from 2000 until midnight, when the skies started to clear and the stars shone brightly all around.

The thick, chilling fog that closed around us as we left Halifax reduced our world to a damp, gray circle no more than fifty yards across. Once we were clear of the land it lifted, but the sea was lumpy and the sky overcast. Shearwaters glided quickly past us on stiff wings, eyeing us coldly, and stormy petrels fluttered over the surface, dabbling their feet in the water, taking no notice of us at all. The spray flew back, wetting our faces and stinging our eyes, as we butted, close-hauled, through the short, steep waves, out into the Atlantic.

Coming off watch on the first night out, I curled up at the far end of my bunk trying to keep clear of the drips coming through the deck just above me. This is no fun, I reflected—no fun at all. I am cold and scared, and there are nearly three thousand miles of ocean ahead of us. What the hell am I doing here? Why did I agree to come on this trip? I kept thinking of all the comforts I had left behind, especially warmth and dry clothes. The leaks would eventually stop—more or less—but only when the sun-baked teak was thoroughly soaked and the seams had tightened up.

The following day we crossed the edge of the continental shelf southeast of Sable Island, a menacing sliver of sand and grass that lies right in the shipping lanes—the scene of countless wrecks. Perhaps the first of these to be recorded (by the great compiler of accounts of early voyages, Hakluyt)[1] occurred in August 1583, when one of the vessels in a small squadron led by the Elizabethan adventurer Sir Humphrey Gilbert ran aground there and broke up. Her crew suffered terribly from thirst and hunger before they reached the coast of Nova Scotia in an open boat. There they were rescued by Basque fishermen (who were by then routinely taking cod from the Grand Banks), and they eventually managed to get back to Europe—unlike Gilbert, who, having coolly turned his back on his shipwrecked companions, was lost on the return voyage.

Sable Island is a desolate spot, inhabited only by wild ponies and sometimes a few research scientists, with a lighthouse at either end. Colin—who relished a difficult pilotage challenge*—had been planning to land there, but even he did not feel like making the detour in these conditions. I was quietly relieved. The first accurate chart of Sable Island was published in 1779 by Cook's mentor, DesBarres. From 1763 to 1773 DesBarres charted the heavily indented coast of Nova Scotia, and he devoted two summers to Sable Island alone. Apparently the surf often broke "mast high" on the two sandbars at either end of the main island, which were "strewn with wrecks for seven leagues" (twenty-one miles).[2] It was difficult and dangerous work, but of vital importance, and DesBarres's chart was one of the first to show longitudes (a word that Colin taught me to pronounce—naval fashion—with a soft "g") based on the Greenwich meridian.

We had now reached the real ocean, where the seabed plunges from a few hundred feet in depth to ten or twenty thousand. The "abyssal plain" stretched out gloomily beneath us. The Canadian Navy had promised us that no icebergs would drift this far south, so the only likely hazards from now on—apart from bad weather—were other vessels. Nevertheless, it was disconcerting to know that we were suspended over several vertical miles of water, kept afloat only by an inch or two of wood.

Standing watch alone on the second night out, I was confronted by an overwhelming sight: half the visible universe, velvet black from horizon to horizon, filled with the brightest stars I had ever seen. Their brilliance was undimmed by the orange glow of man-made light that veils the skies over so much of the land, and they seemed infinitely numerous. Three stars,

---

* He once took a yacht to Mont St. Michel on the north coast of France, where the huge tides come in across the shallow mud flats at a dangerously fast rate, and on the same trip planted a flag on the central islet of the extensive and menacing nearby reef known as Les Minquiers. These are not exploits for the faint-hearted or inexperienced sailor.

named by the old Arab astronomers, formed a brilliant triangle above us, with the Milky Way—a glowing river of light—running among them: Vega, the falling eagle; Deneb, the tail of the hen; and Altair, the flying eagle.[3] The light from the nearest of them had taken sixteen years to reach my eyes. Very occasionally, a shooting star slid soundlessly across the blackness, momentarily animating a spectacle of timeless grandeur and serenity.

As we sailed into the Atlantic, leaving the land farther and farther astern, I watched the night sky as I had never done before. I recognized some of the main constellations and, with the help of a star chart, picked out a few of the fifty-odd "navigational stars" whose coordinates are listed in the *Nautical Almanac*. Aldebaran, Alkaid, Alioth, Antares, Arcturus, Capella, Mirfak—as well as the three stars of the so-called Summer Triangle—were all in sight. The planet Jupiter shone brightly, low in the southeast, while Mars rose later, to be followed shortly before dawn by Saturn. On Colin's advice I was reading Mary Blewitt's classic introduction to celestial navigation for yachtsmen,[4] and a solitary four-hour watch gave me time to watch how the heavens moved. Above me the entire night sky, with all its stars, was slowly revolving counterclockwise around a stationary Polaris. Stars lying closest to the northern celestial pole (the point vertically above the geographical north pole, or in its zenith) never touched the horizon, while the others rose at different points along the eastern horizon, just like the sun by day, climbing up in long majestic arcs before declining slowly in the west. The poet Homer had closely observed the same phenomenon early in the first millennium BCE, as this passage from *The Odyssey* reveals:

> [*Calypso*] *gave him a warm, fair wind, and Odysseus joyfully spread his sail before it, while he sat and steered skillfully with the stern oar. He never allowed sleep to close*

> *his eyes, but kept them fixed on the Pleiades, on Arcturus that sets so late, and on the Great Bear . . . which revolves, keeping watch on Orion, alone never dipping into the stream of Okeanos—for Calypso had told him to keep it on his left.*[5]

I knew, of course, that this celestial motion was in a sense unreal—that it was the earth that turned while the distant stars maintained their imperturbable stillness—but the illusion is too powerful to resist. The earth remains firmly at the center of the sailor's universe, just as it did for the Greek astronomer Ptolemy in the second century CE. It is easy to understand how difficult it was for people of the sixteenth and seventeenth centuries to adapt to the heliocentric view of the universe, and for the purposes of the working navigator the Copernican Revolution might never have occurred.

> *Day 3: Came on watch at 0400. A fabulous dawn with scarcely a cloud in sight. On a close reach under full main and genoa making 5–6 knots. The seas were much calmer and there was a striking change of color. Instead of a dull green, the water is now an amazing sparkling azure. Also much warmer—the Gulf Stream. Alexa is feeling better. The sun shone brightly all day and we had bread and cheese for lunch with beer.*
>
> *Took a meridian altitude with the sextant—my first. Latitude 43°17′ N.*
>
> *Colin cooked another excellent stew for supper and we all had some whisky. Everyone cheerful.*

Though I was very sleepy, the brilliant light and warmth instantly lifted my spirits. The wind had shifted into the south and we were sailing fast, surging smoothly across the waves rather than crashing through them. At last the decks were dry and we could do without oilskins.

We had crossed the "Cold Wall," which marks the division between the frigid, soupy Labrador Current that pours southward out of the Arctic and the vast body of crystalline, lapis-lazuli-blue water that surges out of the Gulf of Mexico between Cuba and Florida and sweeps northward off the east coast of the United States. The volume of the Gulf Stream is so huge that it retains its separate character until well out into the Atlantic, and its vital warming influence is felt across the whole of northern Europe.

A hundred yards ahead of us, dozens of shearwaters were diving on a shoal of fish, and suddenly the surface of the water exploded in their midst: for a startled moment I thought a missile had been launched from a submarine. An enormous streamlined shape emerged, rising at least ten feet in the air. Turning and catching the sun, its flank flashed silver, before it crashed clumsily back into the sea in a colossal shower of spray. I had never seen such an enormous fish—it was a tuna, perhaps half a ton in weight, a seaborne sprinter that could keep pace with a cheetah. The astonishing spectacle lasted only a couple of seconds. Soon the small fry that had attracted the predators had been consumed, and all was still again.

As the sun approached its highest point above the southern horizon, Colin appeared in the main hatch, his silver hair sticking out in all directions, wearing an old guernsey jumper that was full of holes. In his hands was the sextant that had, until now, lain unused in the cabin down below, firmly secured in its square wooden box. Before handing it to me Colin warned me with unusual solemnity never, ever to drop it. "Care of the sextant" was a serious matter: it was a precision instrument and our lives depended on its accuracy. My first lesson in celestial navigation was about to start.

There are many kinds of sextant, and they come in many different sizes—from pocket ones just a few inches across to heavyweight models on a much grander scale. And many materials

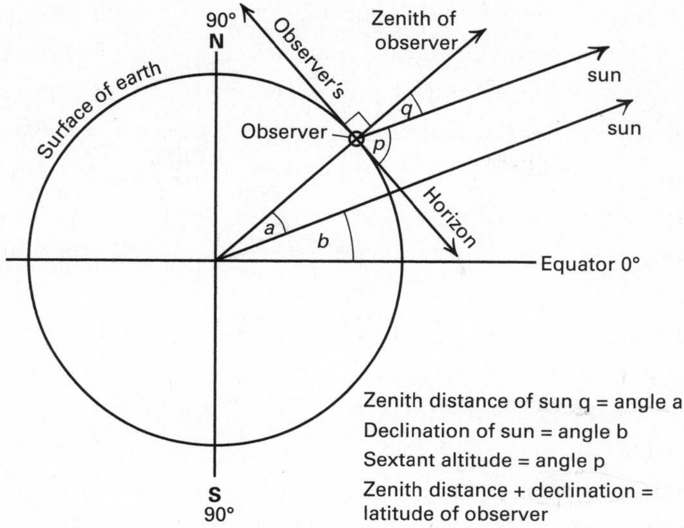

Fig 1: Principles of the Meridian Altitude.

have been employed in making them, from brass and steel to plastic and even cardboard. The essential design, however, has varied little since the eighteenth century, and a good sextant has the reassuring heft and feel of something really well made. With familiarity comes the recognition that this is an instrument perfectly adapted to its purpose: a solution to a practical problem so elegant and efficient as to be quite simply beautiful. But although I had studied a diagram, the sextant now in my hands was bafflingly unfamiliar. Attached to a triangular black steel frame with a wooden handle on one side were two mirrors, several dark shades, a small telescope, and an index arm with a micrometer drum that swung along a silvered arc marked in degrees. Colin showed me how to hold it, with the handle in my right hand and my eye to the telescope.

I had to measure the height of the sun above the horizon just as it reached its highest point in the sky due south of us—as it crossed our meridian. Colin first adjusted the shades on the

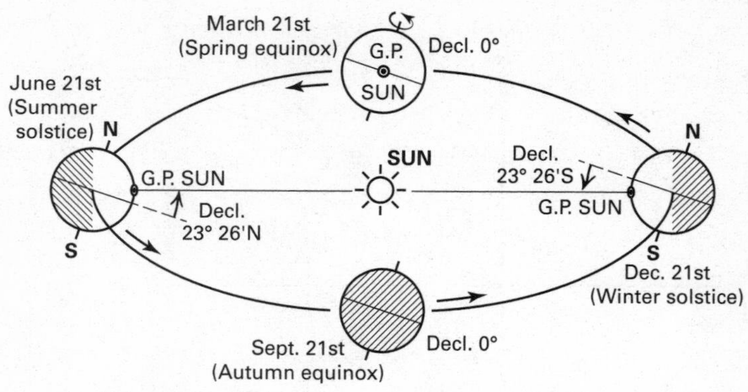

Fig 2: Diagram illustrating the sun's varying declination. (G.P. is the geographical position: see Glossary.)

sextant, then, looking through the telescope, moved the index arm until the sun's lower edge (or "limb," in astronomical jargon) was more or less on the horizon. I then took his place in the main hatch, braced against the slow roll of the boat, and, gripping the handle firmly, peered tentatively through the telescope at the southern horizon.

All I could see at first was a circle divided vertically between a light half and a dark: the left-hand side was the direct view through the plain glass side of the horizon mirror, and the darker right-hand side was the reflected view of the sky above us through the heavily shaded index mirror. Then I found the horizon and, scanning to left and right, caught a glimpse of a brilliant white disc floating just above the dark line of the sea. In a moment it had gone, but then I caught and held it, fascinated to see it moving steadily upward, the gap between the disc and the horizon widening all the time. It was the sun, and I was watching the earth turn.

If I rocked the sextant from side to side, the sun swung in an arc across the sky. By adjusting the micrometer, I brought the disc slowly down until, when the sextant was held vertically, its lower limb was just kissing the horizon. The sun was still moving upward, but much more slowly now as it neared our meridian. After a minute or two, the white disc paused at the top of its arc. Taking the sextant away from my eye, I looked at the scale and read off the angle: 64° on the main scale and 41 (60 minutes to one degree) on the micrometer. This was the sun's meridian altitude, or "mer alt."

Colin took the sacred instrument from me and confirmed the reading. I looked up the sun's declination in the *Nautical Almanac* and made a few corrections to the observed angle. In a few minutes, to my astonished delight, I completed the simple addition and subtraction sums that yielded our latitude.[6] We were somewhere on the parallel of 43°17' North, and—as Colin observed—I was now as well equipped to find my way safely across an ocean as any European mariner before the time of Captain Cook.

Chapter 3

The Origins of the Sextant

Day 4: Woken at 0400 and watched a perfect sunrise at 0535. Still reaching at a good 5 to 6 knots on course of about 110°. Have covered 330 miles. Beginning to feel very grubby but there's no fresh water to spare for washing.

Another hot, clear, calm day with wind SSW force 2–3. Passed a bulk cargo ship going the other way. Started reading Slocum[1] sitting in the sun, then took a nap from 10–12. Then did another mer alt—42°58' N.

Our course—approximately 120° magnetic—is meant to take us clear of the Tail of the Bank, where there are likely to be many fishing boats. Slocum sailed on just this route when he set off on his round-the-world voyage.*

Alexa saw some dolphins. I could hear them down below. Everyone in very good spirits.

The heavens have always fascinated people, and we have long looked to them for guidance, though we were not the first animals to do so. Many different species use the sun, moon, and stars to help them reach their destinations—whether these are nests a few yards away, or breeding grounds on the other side of the world. The magnificent monarch butterfly, for example, relies on an internal sun compass to find its way at the end of every summer from the eastern United States south to the mountains

---

* The southern tip of the Grand Banks that extend to the south of Newfoundland—once the home of the great cod fishery.

of central Mexico, where vast numbers pass the winter clinging to the trees. On a more modest scale, dung beetles have recently been shown to use the orientation of the Milky Way to help them roll dung balls back to their nests by the shortest route,[2] and honeybees use polarized sunlight to navigate to and from their hives on foraging trips.[3] Mystery still surrounds the exact nature of the homing pigeon's skills, but they seem to involve a magnetic sense, coupled with a kind of sun compass, and the ability to hear low-frequency sound, such as that produced by the breaking waves that mark the line of the coast.[4] Some migrating birds rely on Polaris, and seals too can steer by the stars.[5]

Perhaps our pre-human ancestors wondered about the sun, moon, and stars before our own species appeared a couple of hundred thousand years ago. Certainly the earliest humans must have realized that most "heavenly bodies" (the old term is irresistible) moved in regular and predictable ways, and that these motions were linked to vitally important seasonal variations in the activities of plants and animals, as well as the length of the days, the weather, and the tides. The structures left behind by our prehistoric ancestors present many puzzles but they do at least reveal that their builders had an excellent grasp of the motions of the sun and moon. At dawn on the shortest day of the year (the winter solstice, when the sun stands vertically above the Tropic of Capricorn), the first light still strikes through a carefully placed stone aperture above the entrance to the great passage grave at Newgrange, in the Boyne valley in Ireland. Shooting down a long, low corridor of massive stones it briefly illuminates the burial chamber at the heart of the huge mound. Stonehenge may be a little younger than Newgrange—perhaps only 4,500 years old, rather than 5,200—but the behavior of the sun and moon clearly mattered deeply to the designers of this elaborate array of standing stones. The sun on the longest day of the year (the summer solstice, when the sun stands vertically over the Tropic of Cancer), when observed from the center of the stone circle, rises just over the top of a single, lonely stone at the perimeter (the Heelstone), as does the full moon closest to the winter solstice.[6]

More recent than these Neolithic monuments, a mere 3,600 years old, is the spectacular Nebra Sky Disc. Illegally excavated in Germany in 1999, and retrieved after an undercover police operation, it seemed at first too good to be true. Many experts feared that the dinner-plate-sized object was a fake, but extensive tests have shown that it is genuine. It is perhaps the oldest known visual representation of the cosmos, revealing for the first time that Bronze Age Europeans—like the supposedly more sophisticated inhabitants of ancient Egypt and Mesopotamia—paid close attention not only to the sun and moon, but also the stars. The tight group of seven stars represent the Pleiades as they appeared at that epoch, and the Disc may have been used to harmonize the solar and lunar calendars, a process hitherto thought to have been a Babylonian discovery a thousand years after the disc was made.

Accurate calendars would, of course, have been useful for many purposes, such as choosing when best to plant crops, but it is hard to believe that this was the Nebra Sky Disc's only purpose. Creation myths from around the world offer wildly varied accounts of the origins of the sun, moon, and stars and the significance of their behavior. The extraordinary imaginative energy they display plainly arose from deep concerns about our place in the universe and the meaning of life and death. Such concerns must surely also have influenced the designer of the disc. It has been suggested that the curved piece of gold at the bottom of the disc represents a boat—perhaps one that safely carries the sun across the ocean after it has set. It might equally refer to the passage of the soul to the afterlife. We cannot help sensing that this extraordinary object, like the many prehis-

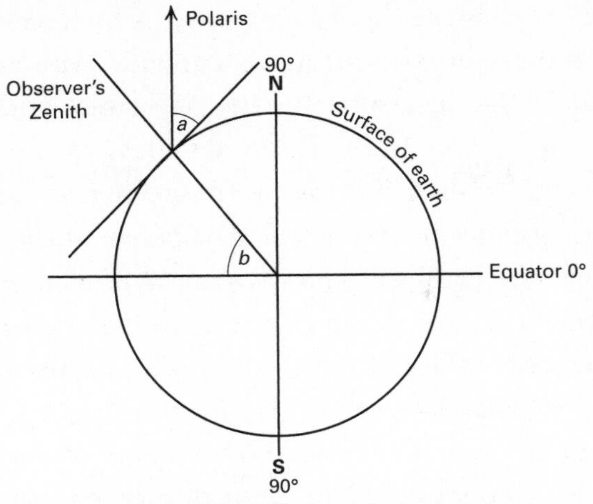

Fig 3: Diagram illustrating the equivalence of Polaris altitude and latitude.

toric structures that are aligned with the heavens, embodies profound, if mysterious, spiritual beliefs.

Until very recent times the heavens shaped the patterns of everyday life. The farmer judged when to sow his crops by looking at the night sky, and the sun and stars told him the time of day, long before the first mechanical clocks were invented. One of the western portals of the great thirteenth-century cathedral of Amiens in northern France is decorated with the signs of the zodiac, each one accompanied by a depiction of the activities appropriate to the month with which it was associated—such as sowing, reaping, cutting hay, and treading grapes. Similar motifs appear on many other medieval buildings. But people did not rely on the heavens only to plan their communal activities; they also thought that the sun, moon, and stars could foretell what lay in store for them as individuals or nations. This belief was—and is—so widespread as to qualify as a cultural universal.

The skies were, of course, especially important for sailors. The moon enabled them to predict the height of the tide;

"full" and "new" bring the "spring" tides, which have the widest range and produce the strongest currents, while the "half" moon signals the "neaps," with the narrowest and weakest. By day the sun, rising in the east and setting in the west, told mariners roughly which way they were steering, as did Polaris—much more simply—by night. For navigators in the Northern Hemisphere, the height of Polaris was the crucial measure of latitude—the only one, in fact, until astronomers were able to produce accurate tables of the sun's varying declination at the end of the fifteenth century.

As this diagram makes clear, the height of Polaris above the northern horizon is equivalent to the observer's latitude. Thus if Polaris is vertically overhead, you must be at the geographical North Pole, while if it appears right on the horizon you must be on the equator. In this diagram, its height above the horizon is 50 degrees and it follows that you must be 50 degrees north of the equator—in other words your latitude is 50 degrees North.*

So vital was Polaris to the seafarer that the English simply called it "the Star," and Shakespeare knew enough to press it into service in Sonnet 116:

Let me not to the marriage of true minds
Admit impediments. Love is not love
Which alters when it alteration finds,
Or bends with the remover to remove:

O, no! it is an ever-fixèd mark,
That looks on tempests and is never shaken;
It is the star to every wandering bark,
Whose worth's unknown, although his height be taken. . .†

---

* A small adjustment has to be made to allow for the fact that Polaris is not quite vertically above the North Pole.

† Shakespeare is clearly referring to Polaris, but the last line of the quotation is puzzling. A plausible interpretation is that the "worth" of the star is unknown to

In medieval Latin, Polaris was Stella Maris—"star of the sea"—a term that was applied also to the Virgin Mary, whose sky-blue cloak is emblazoned with a star in many early European paintings. The theologian Alexander Neckham (1157–1217) likened Mary to the Pole Star standing at "the fixed hinge of the turning sky" and by which the sailor at night directs his course. Polaris must have seemed a perfect symbol of the Mother of God, the immaculate spiritual guide and intercessor. The name Stella Maris was sometimes also applied to the ship's steering compass on which the thirty-two "points" are still often marked in the form of a star.[7] Even today Stella Maris is a common name for fishing boats and "true north"—solidly reliable, unlike its variable magnetic cousin—was marked on old charts with a star. (True north marks the direction of the geographical north pole, which is fixed,* whereas the magnetic north pole—like its southern counterpart—wanders and is at present several hundred miles distant from it.)

BEFORE THE END of the thirteenth century, the Venetian explorer Marco Polo recorded that he had measured the altitude of Polaris on the coast of India, though how he did so is unclear. The earliest references to the measurement of the height of Polaris by Portuguese navigators date from the mid-fifteenth century, but it seems unlikely that they were using purpose-built instruments to make their observations.[8] In theory they could have used the astronomer's astrolabe, an elaborate device that permitted the observer to find the time of day, as well as solv-

the lovers who are the crew of a "wandering bark"; their inconstancy prevents them appreciating the true value of the star's "height" even when they have "taken" it. They can measure its height but do not know how to deduce their latitude from it.

* This is true for almost all practical purposes, but movements of the earth's crust mean that its geographical position may be subject to small changes.

ing other astronomical problems. It had by then achieved a high level of sophistication but it is doubtful that many ordinary sailors would have known how to operate one, and such a complicated and costly instrument would not have been required merely for the purpose of measuring the altitudes of heavenly bodies.

A radically simplified version, known as the mariner's astrolabe, was, however, widely adopted for use at sea during the sixteenth century,[9] by which time solar declination tables were available. It was a metal disc or hollow circle with a scale of degrees engraved on its circumference. While it was suspended from its ring, the navigator would adjust the "alidade" (a revolving bar with peephole sights) so that it was aligned either with the sun or with Polaris and its height could then be read off from the scale. The motion of the ship, however, and the effects of the wind limited the usefulness of any instrument that had to be freely suspended, and it would have been much easier to obtain accurate readings from it on dry land.

The Portuguese poet Luís Vaz de Camões (*c.* 1524–80), who sailed out to India in 1553, witnessed the use of an astrolabe at first hand—though it is not clear which kind—after his ship anchored off the west African coast. In his great epic *The Lusiads* (first published in 1572), he describes the scene:

Like clouds, the mountains we spied
Began to reveal themselves;
The heavy anchors were readied;
Now arrived, we took in the sails.
And so that we would better know
Where we were in these remote parts,
Using that new instrument, the astrolabe,
An invention of subtle skill and wisdom.

We then landed on an open shore
Where the crew scattered, wishing
To see strange things in the land
Where no other people had trod.
But I, with the pilots, on the sandy beach
To find out where I was,
Remained to take the height of the sun
And measure the painted universe.

We reckoned that we had already passed
The great circle of the Tropic of Capricorn
Being between it and the frozen southern
Circle, the most secret part of the world. . . .[10]

The seaman's quadrant, an even simpler device that was probably in use at a much earlier date than the mariner's astrolabe, relied on a plumb-line, so it too would have been of limited util-

Fig 4: Back-staff (left) *and cross-staff* (right).

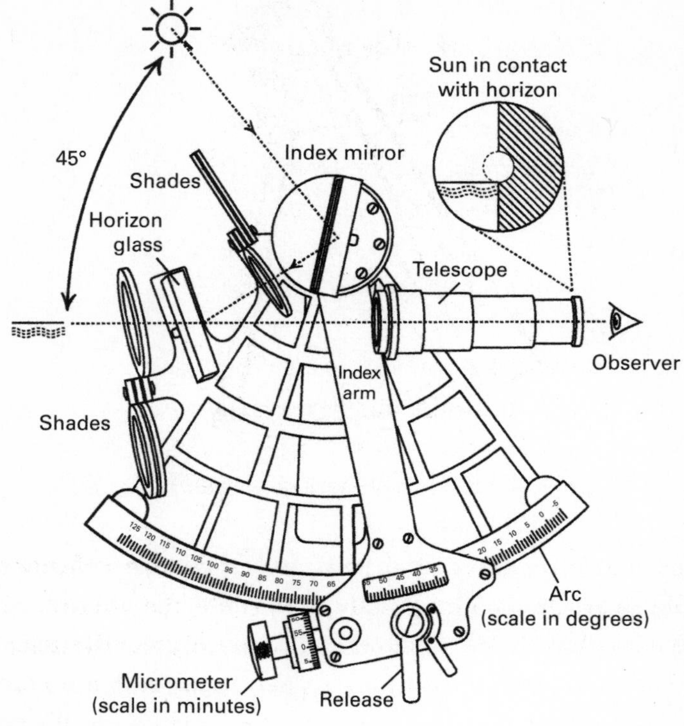

Fig 5: Diagram of a sextant, showing the double-reflection principle.

ity on the open sea. Later developments included the cross-staff (first mentioned in 1545)[11] and the more sophisticated back-staff (invented by the English navigator John Davis in the late sixteenth century). These were much more practical than the astrolabe and quadrant, but they were hardly precision instruments. In the latter part of the seventeenth century, as the scientific revolution fast gathered momentum, astronomers started to explore ways in which more accurate sights could be obtained from the moving deck of a ship. It was from this process that the sextant eventually emerged.

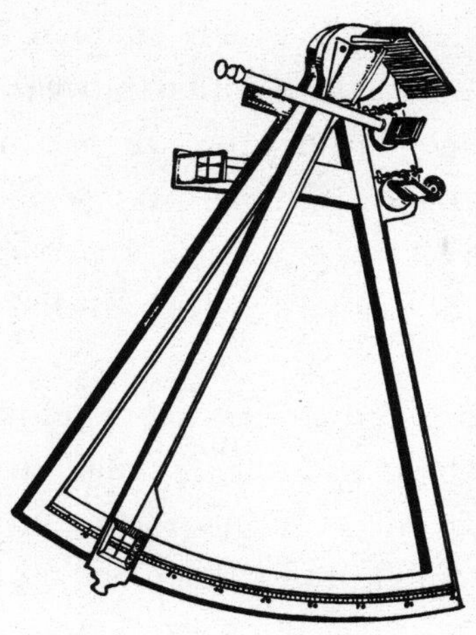

Fig 6: Hadley's reflecting quadrant.

The sextant employs the principle of double reflection to enable the user simultaneously to observe the horizon and a chosen heavenly body, and to measure the angular distance between them with great accuracy. Its key virtue is that it marries the two in a single, steady image that is completely unaffected by the movement of the observer (or the deck on which he or she is standing) so long as the instrument is kept pointing in the right direction. The sextant is also versatile. Unlike the astrolabe or quadrant it can be used for measuring angles in *any* plane—for example, between two heavenly bodies, or between two objects on the surface of the earth.

The sextant was the offspring of an earlier invention, the so-called reflecting quadrant. Sir Isaac Newton can take credit for designing the first device of this kind, plans for which were shown to the Royal Society in 1699.[12] Another Fellow of the Royal Society, John Hadley (1682–1744), came up with two designs, similar to Newton's though apparently not derived from them, which he presented to that institution in

May 1731.[13] One of these was widely adopted following successful sea trials conducted the following year by the Oxford professor of astronomy John Bradley, who was later to become Astronomer Royal.[14] By one of those strange coincidences that seem common in the history of science, an American—Thomas Godfrey—independently came up with a similar design almost simultaneously.[15]

Confusingly, the reflecting quadrant is actually an octant—its arc is one-eighth of a circle (45 degrees) rather than one-quarter. It is capable of measuring angles up to 90 degrees, thanks to the double-reflection principle. The invention of "Hadley's quadrant" marked a revolution in the history of marine navigation. For the first time, it was possible to measure the altitudes of heavenly bodies from the moving deck of a ship at sea with great precision—in fact, with the help of a vernier scale, quadrants permitted readings to the nearest minute of arc (one-sixtieth of a degree).[16] Its larger cousin, the sextant, which was later to become the instrument of choice for accurate offshore navigation, could measure angles up to 120 degrees. Both the quadrant and the sextant were far superior to any instruments previously available for measuring angles at sea, in terms of both precision and ease of use. The sextant's original design was so perfect that it has to this day remained essentially unchanged, and with its introduction the solution of the greatest problem of celestial navigation—the accurate determination of longitude at sea—at last became a practical possibility.

HAVING LEARNED HOW to do a mer alt, I could, in theory, have found my way home to England simply by following the right parallel of latitude: 49°30' North brings you nicely into the English Channel—halfway between Ushant and Scilly.* Mariners relied entirely on "latitude sailing" of this kind for hundreds of years before the longitude problem was solved, but it is subject to one potentially disastrous drawback: unless you know how far east or west you have travelled, and the coordinates of your destination, you cannot be sure when you are going to arrive. Latitude sailing also suffers the disadvantage that the shortest distance between two points on the surface of a sphere is a "Great Circle" (a circle whose center coincides with the center of the earth), not a parallel of latitude.† The difference can be significant if the voyage is a long one.

You might suppose that it would be a simple matter for the sailor to work out his position in mid-ocean just by measuring the distance he has travelled on a particular course. This is known as "dead reckoning" (DR), a term that could well have been chosen by someone with a black sense of humor, though

---

* The words "Ushant and Scilly" remind me of the classic sailor's song "Spanish Ladies"—one of Colin's favorites—which incidentally sheds some light on old-fashioned navigation:

Farewell and adieu to you fair Spanish ladies
Adieu and farewell to you ladies of Spain,
For we've received orders for to sail to old England
And hope in a short time to see you again.

We'll rant and we'll roar like true British sailors
We'll rant and we'll roar o'er all the salt seas
Until we strike soundings in the Channel of old England
From Ushant to Scilly is thirty-five leagues.

We hove our ship to with the wind in the southwest
We hove our ship to, boys, fresh soundings to take.
It was forty-five fathoms and a fine sandy bottom
So we squared our main yard and up-Channel did make.

† Unless the two points both happen to lie on the equator, which is itself a Great Circle.

it is actually said to derive from "deduced reckoning." Even today, when it is possible to measure speed and distance travelled through the water with great accuracy, DR is an imprecise science. Many factors affect a vessel's rate of progress "over the ground" (that is, relative to the seabed), some of which are very hard to assess. Ocean currents are one example. These can be strong, but they are fickle, seldom running steadily in one direction or with a constant strength. Steering an accurate course is also much trickier than the landsman might suppose: it is nothing like driving a car down a road. Leaving aside human error, and the tendency of sailing vessels to "sag" (drifting sideways—or making "leeway"—when heading to windward, or "close-hauled"), the magnetic compass itself is subject to large errors—which were not well understood until the nineteenth century. Unless correct allowances are made for all these effects, the navigator will soon be lost. DR is, in practice, highly unreliable and especially so over long distances because the errors tend to accumulate—as Álvaro de Mendaña's experiences in the sixteenth century vividly showed.

There is an extraordinary passage in *Moby-Dick* where Herman Melville contrasts the reliability of celestial navigation with the uncertainties of DR in order to dramatize Captain Ahab's descent into madness. Consumed by hatred for the white whale that has cost him his leg, Ahab takes his last mer alt seated in the bows of one of the open whaleboats in which he hopes to hunt it down.

> *At length the desired observation was taken; and with his pencil upon his ivory leg, Ahab soon calculated what his latitude must be at that precise instant. Then falling into a moment's revery, he again looked up towards the sun and murmured to himself: "Thou sea-mark! Thou high and mighty Pilot! Thou tellest me truly where I am—but canst*

> *thou cast the least hint where I shall be? Or canst thou tell where some other thing besides me is this moment living? Where is Moby Dick?"*

Ahab gazes thoughtfully at the quadrant, handling its "numerous cabalistical contrivances" one after another, and then mutters:

> *"Foolish toy! babies' plaything of haughty Admirals, and Commodores, and Captains: the world brags of thee, of thy cunning and might; but what after all canst thou do, but tell the poor, pitiful point, where thou happenest to be on this wide planet, and the hand that holds thee: no! not one jot more! Thou canst not tell where one drop of water or one grain of sand will be to-morrow noon: and yet with thy impotence thou insultest the sun! Science! Curse thee, thou vain toy. . . . Curse thee, thou quadrant!"*

To the astonishment of his crew, Ahab then dashes the instrument to the deck:

> *"no longer will I guide my earthly way by thee: the level ship's compass, and the level dead-reckoning, by log and by line: these shall conduct me, and show my place on the sea. Aye," lighting from the boat to the deck, "thus I trample on thee, thou paltry thing that feebly pointest on high; thus I split and destroy thee!"* [17]

Chapter 4

Bligh's Boat Journey

Day 5: Took the 0400 watch again. Another brilliant day with southerly force 2–3 wind, occasionally 4. Scarcely any cloud except on the southern horizon, where there always seems to be a patch of cumulus.

After breakfast we checked our DR, which puts us somewhere near the Tail of the Bank. I did another mer alt and Colin plotted our exact position using an earlier timed sun sight—latitude 42°42′ N, longitude 52°13′ W. Still on a course of 120° at about 5 knots.*

Over supper I mentioned how I had first heard of the sextant when I saw *Mutiny on the Bounty*. This triggered a string of reminiscences from Colin, who recalled the mutiny that broke out in 1931 at Invergordon in Scotland aboard some of the Royal Navy's greatest ships—including the famous battle cruiser HMS *Hood*. Pay cuts were blamed at the time, but low morale on the big ships was the main factor, he thought. The smaller, more tightly knit crews of destroyers and frigates had experienced fewer problems. I asked what he thought of Bligh. Colin did not think the film had painted a fair portrait of him: Bligh had been a great seaman and navigator and, like Cook, had risen from the ranks. Maybe his explosive temper reflected some kind of social insecurity. Colin also objected that, since Bligh was in his

---

* The method is explained in Chapter 15.

mid-thirties when he commanded the *Bounty*, Trevor Howard had been far too old to play him.

Colin was right about Bligh's skills as a navigator. Bligh had sailed with Cook as master of the *Resolution*, a post to which he was appointed at the unusually early age of twenty-one. He seems to have enjoyed Cook's approval; he certainly demonstrated great skill as a surveyor and draftsman. But he was a difficult man. J. C. Beaglehole, in his magisterial life of Cook, says that he "saw fools about him too easily," and that even at this early stage in his career he displayed "the thin-skinned vanity" that was always to be his curse: "Bligh learnt a good deal from Cook: he never learnt that you do not make friends of men by insulting them."[1]

Bligh was actually involved in not one but three mutinies. These tempestuous events did not stop him reaching the rank of vice admiral,* but they have overshadowed his substantial achievements. Of these the most remarkable was his voyage in an overloaded twenty-three-foot open boat after being set adrift by the *Bounty* mutineers in the Tonga Islands. The mutineers, led by the master's mate, Fletcher Christian, comprised more than half the *Bounty*'s crew, and they were—in Bligh's words—"the most able men of the ship's company."[2] Bligh later speculated that the temptations of Tahiti were the main cause of the mutiny. The crew had just spent twenty-three lazy weeks there while the gardener prepared the breadfruit seedlings for transplantation to the West Indies, and discipline had inevitably suffered.

Bligh was not surprised that "a set of sailors, most of them void of connections," should wish to "fix themselves in the midst of plenty on one of the finest islands of the world, where they

---

* This did not bring any special credit to Bligh. Promotion beyond the rank of post-captain was simply a matter of seniority: anyone could reach "flag rank" if they lived long enough.

need not labour, and where the allurements of dissipation are beyond anything that can be conceived." However, he claimed to be aware of no discontent and bitterly complained that he had thought himself to be on the friendliest terms with Christian. So he felt not only shock but also a personal sense of betrayal when, just before sunrise on April 28, 1789, Christian, accompanied by three other men, came into Bligh's cabin, tied his hands behind his back, and threatened him "with instant death" if he made the least noise. While Christian held a bayonet to his throat, the members of the crew who had refused to join the mutiny were put over the ship's side into the launch. The captain's clerk tried to save Bligh's surveys and drawings, but was forbidden to do so. Nor was Bligh allowed to take the chronometer or any charts. At last he himself was forced to board the open boat, which was promptly cast adrift. Equipped only with a sextant and compass,[3] and very limited supplies of bread, pork, water, rum, and wine, Bligh now faced the almost overwhelming challenge of bringing to safety the eighteen men who accompanied him.[4]

Bligh decided first to lay in a supply of breadfruit and water at the nearby island of Tofoa (now Tofua), but this plan went badly wrong. They were able to obtain very little in the way of provisions, and the natives—some of whom recalled Bligh from his visit to the Tongan archipelago with Cook fifteen years earlier—turned hostile when they realized the sailors were poorly armed and quite alone. Eventually they gathered on the beach, menacingly knocking stones together, and Bligh—who had witnessed Cook's death—saw that an attack was imminent. He ordered all his men to get aboard the boat as quickly as possible, but stones began to fly and a member of the crew who had run back up the beach to cast off was clubbed to death. Bligh cut the painter and they escaped, under a barrage of well-aimed missiles, leaving their unfortunate comrade behind.

Despite the desperate shortage of supplies, Bligh and his companions decided not to risk landing on any of the neighboring islands. Instead they headed west for Timor, in the Dutch East Indies, some 3,600 nautical miles away, as it was the nearest place where they could be sure to find help—and report the mutiny. To give some sense of the scale of this voyage, that is roughly the distance from Land's End to the northeast coast of Brazil.

Shortly after leaving Tofua they were caught in a heavy gale:

> *The sea ran very high, so that between the seas the sail was becalmed, and when on the top of the sea it was too much to have set: but we could not venture to take in the sail for we were in very imminent danger and distress, the sea curling over the stern of the boat, which obliged us to bail with all our might. A situation more distressing has perhaps seldom been experienced.*[5]

Everything now depended on Bligh's exceptional navigational skills and remarkable memory. Taking observations with the sextant to determine their latitude—a difficult feat in an overcrowded boat often tossed about in heavy seas—while keeping track of their westerly progress with the help of a makeshift log-line,[6]* Bligh sailed toward the Great Barrier Reef, setting their course in accordance with his apparently detailed recollection of the charts he had been forced to leave behind. The food and water now had to be very strictly rationed. Such was Bligh's devotion to duty that, even in these desperate circumstances, he continued to keep careful notes of the islands they passed—

---

* A length of rope was carefully knotted for this purpose and attached to a small wooden board. Bligh taught several members of the crew how to count seconds "with some degree of exactness" in order to time how long it took to run out when the wooden "log board" was thrown over the side.

including the Fiji group, which they were the first Europeans to discover—recording their latitudes and estimating their longitudes as best he could. In addition to their growing hunger and thirst, the lack of space made life on board the boat "very miserable." Bligh kept half the crew sitting up on watch while the other half lay down in the bottom, or on the chest in which they kept their small supply of bread:

> *Our limbs were dreadfully cramped, for we could not stretch them out; and the nights so cold, and we so constantly wet, that, after a few hours sleep, we could scarce move.*[7]

After three weeks things were starting to look hopeless. An occasional teaspoon or two of rum or wine helped to keep their spirits up, and the redoubtable Bligh survived almost without sleep. Another gale brought them to the brink of disaster, but even then Bligh was still taking sights:

> *At noon it blew very hard, and the foam of the sea kept running over our stern and quarters; I however got propped up, and made an observation of the latitude, in 14°17′ S; course N 85° W, distance 130 miles; longitude made 29°38′ W.*
>
> *The misery we suffered this night exceeded the preceding. The sea flew at us with great force, and kept us bailing with horror and anxiety. . . . At dawn of day I found everyone in a most distressed condition, and I began to fear that another such night would put an end to the lives of several, who seemed no longer to support their sufferings.*[8]

When the weather eventually improved, the heat of the sun became a serious problem, but the appearance of large numbers of birds, and the sight of stationary clouds on the western horizon, at last suggested that they were approaching land. In the middle of the night the helmsman heard the sound of breakers, and Bligh woke to see them "close under our lee, not more than a quarter of a mile distant from us": they had made their landfall on the Great Barrier Reef, just as Bligh had intended. The following day they began to search for a gap through which they could pass:

> *The sea broke furiously over every part. . . . I now found that we were embayed, for we could not lie clear with the sails, the wind having backed against us; and the sea set in so heavy towards the reef, that our situation was become unsafe. We could effect but little with the oars, having scarce the strength to pull them; and I began to apprehend that we should be obliged to attempt pushing over the reef. Even this I did not despair of effecting with success, when happily we discovered a break in the reef. . . .*[9]

Having passed within the reef, Bligh took a mer alt in order to determine the latitude of the channel through which he had just passed—12°51' South—and recorded his DR longitude: 40°10' West of Tofua. In fact the distance is more like 32 degrees, which goes to show just how hard it is to estimate the rate of progress at sea, even for an expert like Bligh. They then headed north, looking for a convenient place to land where they would not be at risk of attack from the natives. They found a suitable island and feasted on oysters and berries, their morale much improved. It now began to look as if they might have a chance of surviving. To be able to sleep ashore was, in Bligh's view, almost as valuable to them as food.

Bligh followed the coast of Cape York Peninsula to the north, and passed through the Torres Strait into the open sea to the west, just as Cook had done in 1770. Given the extraordinary intricacy of the navigation among the many reefs and islands, he dutifully felt he should record directions, and regretted his failure to do so:

> *I . . . think that a ship coming from the southward, will find a fair strait in the latitude of 10° S. I much wished to have ascertained this point; but in our distressful situation, any increase of fatigue, or loss of time, might have been attended with the most fatal consequences. I therefore determined to pass on without delay.*[10]

The remainder of the voyage was, if anything, even more testing than the earlier passage from Tonga. They survived on dried clams, and Bligh managed to catch a booby with his bare hands: he divided the blood among those who were in the worst condition and kept the rest of the bird for the next day. A small dolphinfish later gave them some relief, but the crew were growing steadily weaker, and Bligh began to fear that some of them would not last much longer. The boatswain "very innocently told me that he thought I looked worse than anyone in the boat. The simplicity with which he uttered such an opinion amused me and I returned him a better compliment."

At three in the morning of June 12, 1789, they at last sighted land:

> *It is not possible for me to describe the pleasure which the blessing of the sight of this land diffused among us. It appeared scarcely credible to ourselves that, in an open boat, and so poorly provided, we should have been able to reach the coast of Timor in forty-one days after leaving Tofoa, having by that time run, by our log, a distance of 3,618 miles; and that, notwithstanding our distress, no one should have perished in the voyage.*[11]

Bligh recalled that the Dutch settlement was at the southwest end of the island, so he headed that way, and finally found his way to Cupang (now Kupang), where he landed on June 14. Although only one of the castaways had died (in the native attack on Tofua at the outset of the voyage), Bligh and his crew were a shocking sight, some scarcely able to walk as they struggled ashore:

> *An indifferent spectator would have been at a loss which most to admire: the eyes of famine sparkling at immediate relief, or the horror of their preservers at the sight of so many spectres, whose ghastly countenances, if the cause had been unknown, would rather have excited terror than pity. Our bodies were nothing but skin and bones, our limbs were full of sores, and we were clothed in rags: in this condition, with the tears of joy and gratitude flowing down our cheeks, the people of Timor beheld us with a mixture of horror, surprise and pity.*[12]

Unfortunately, their problems were by no means over. The Dutch East Indies were extremely unhealthy, and the various endemic tropical illnesses, including malaria and dysentery, were to take a heavy toll on Bligh's crew—as they did on so many European visitors. The Dutch governor, himself fatally sick, nevertheless made sure that the castaways were well looked after, and on August 20 Bligh was at last able to take passage to Batavia (modern Jakarta) in a small schooner. On March 14, 1790, he reached Portsmouth with eleven out of the open boat's original crew of nineteen. The remainder had died of illness either in Indonesia or on the homeward voyage.

Fourteen of the mutineers who had decided to settle in Tahiti were hunted down there, and four of them drowned when the ship in which they were being brought home for trial was wrecked off the Great Barrier Reef. The ship's name was *Pan-*

dora and, predictably, the cage in which the unfortunate prisoners were being held came to be known as "Pandora's box." Three of the surviving mutineers were hanged, but another, the young midshipman Peter ("Pip") Heywood, was pardoned and later enjoyed an illustrious career as a marine surveyor.[13] The ringleader, Fletcher Christian, and eight others, together with some men and women from Tahiti, escaped to Pitcairn Island, where they remained unmolested even after they were discovered there in 1808. Their descendants live there today.

To have brought an overladen open boat across nearly four thousand miles of tropical sea, without charts and with grossly inadequate provisions, stands as one of the most remarkable feats in the history of seafaring. Of course luck must have played a part in Bligh's survival and that of his crew, as did their powers of endurance: they were certainly a tough group of men. Bligh's own bloody-minded determination to see the mutineers brought to justice probably helped to keep him going. Were it not for his skill with the sextant and geographical memory, however, they would have had no hope of reaching Cupang. The mutineers no doubt had reason to complain about their commander's behavior, but they fatally underestimated his extraordinary navigational abilities.

Chapter 5

Anson's Ordeals

Day 6: Up again at 0400. Colin says we're very close to where the Titanic *went down. More fabulous weather with wind S force 2–3. Sighted a Sanko Line ship and tried to raise her on the radio-telephone—no luck. More sextant practice.*

After supper Colin talked about his time as gunnery officer on board the battleship Prince of Wales *when she and the battle cruiser* Hood *were in pursuit of the German battleship* Bismarck. *When the battle started* Bismarck *had the "windward station"—this gave her an advantage, just as in the days of sail.* Bismarck's *rangefinders were pointing downwind whereas* Prince of Wales *and* Hood *were plowing into heavy seas that showered theirs with spray. If Colin's guns had found* Bismarck's *range sooner she would have had to alter course and maybe* Hood *would have survived. But* Prince of Wales *did manage to score a crucial hit on* Bismarck.[1] *There were tears in his eyes.*

We also talked about the Battle of Jutland. Colin said that the British commander-in-chief had struggled to determine his exact position as he was closing with the German fleet because poor visibility had prevented any sights being taken. Strange to think that these great warships relied on sextant and chronometer to find their way—just like us.[2]

Saecwen was not much better equipped in navigational terms than the *Bounty*. Like Bligh, we steered by magnetic compass and fixed our position by the sun and stars. It is true we had a

"Walker" log to measure the distance we travelled through the water—a mechanical device that sat on the stern counting the turns of a brass impeller that we trailed behind us. This was more sophisticated than the kind of log that Bligh would have used, but it was still a piece of nineteenth-century technology.

In addition to an old-fashioned lead-line, we had an electronic echo sounder and a radio direction-finder (RDF), both of which would have amazed Bligh. But the echo sounder could measure depths down to a few hundred feet at best, so it was helpful only in coastal waters, and since the marine radio beacons had a range of no more than a couple of hundred miles, the latter too was of little use to us in mid-ocean. Fairly accurate radio-based navigation systems, like LORAN, had been developed during World War II, but the receivers were bulky and expensive and we did not have one. Early forms of satellite navigation were already available but only for military purposes, and GPS was still on the drawing board. We carried a radiotelephone, which turned out to be very temperamental, but apart from flares we had no other means of calling for help in an emergency.

Saecwen herself was ten years old, and handsome without being flashy. She would now be regarded as a "classic" yacht. With a long, deep keel, she was slow and heavy by modern standards, and a bit the worse for wear after her tough outward passage across the Atlantic. We had already had to repair some of her sails, and rust stains were starting to trickle down her white topsides. Most new yachts were already being built of glass fiber-reinforced plastic (GRP), with aluminum masts and stainless-steel rigging and deck fittings, but *Saecwen* was old-fashioned. She was built almost entirely of traditional materials—a teak deck, with a wooden mast and a hull of copper-fastened mahogany planks on oak frames. Beneath the sliding main hatch—at the forward end of the cockpit—a few steps descended into the cabin. The galley, with a small two-burner gas-fired stove and a tiny sink, was immediately on the left, while the chart table with the radiotelephone and RDF set lay on the right.[3] Beyond the galley and the chart table was the saloon, a space perhaps ten feet by eight with a table in the middle and a settee berth on either side. It was lined with lockers—one marked with a red cross for the medical kit—and there were small bookshelves with bars to hold in their contents in heavy weather. Three oblong windows let in light at deck level on either side. Beyond the main bulkhead lay the "heads"—miniature pump-action lavatories—and the fo'c's'l, where there were two more berths and stowage for oilskins and sails. Small electric lights were dotted around the cabin but most of the time we relied on brass oil lamps.

Below deck, *Saecwen* had a very particular smell I can still vividly recall—a musty mixture of damp timbers, diesel oil, paraffin, oilskins, and dirty clothes, coupled with the scent of the ripening fruit and vegetables in the cargo nets overhead. Not very appealing, perhaps, but it was far better than the sharp scent of epoxy resin that never quite vanishes from GRP boats. Being built throughout of wood, *Saecwen* even sounded different from a plastic yacht: footsteps on deck and the thump of waves against the hull were muted and distant, and partly for that reason her white-painted saloon felt especially cozy. The eighteenth-century navigators would have felt quite at home aboard her.

The fast-spinning impeller that trailed at the end of the log-line skipped through our boiling wake as we continued reaching fast to the east under full sail. We passed through patches of yellow Gulf weed, and a sharp dorsal fin slowly zigzagging through the water, like a hound picking up a trail, revealed the presence of a large shark. A half-inflated purplish plastic bag floated by, a depressing reminder of man's polluting habits, until on closer inspection it turned out to be a Portuguese man-of-war—a medusa, trailing its long blue fringe of stinging tendrils.

The fine weather continued and we fell into an easy routine, eating together, taking sights, and otherwise either sleeping or standing watch—four hours on, four hours off at night for Colin and me, with more flexibility during the day, when Alexa helped us out. Every twenty-four hours we recorded good runs of 150 miles or more, the fore hatch stood open and the steady draft of warm air gradually dried out everything down below. On deck everything was now covered with a sparkling rime of salt crystals. We hardly had to touch the sheets, and the self-steering gear[4] kept us on a steady course relative to the wind. Apart from navigation and preparing food, there was little to do apart from keeping a lookout, reading, and occasionally writing up the log.* We kept a close eye on the western sky for any change in the weather—sometimes clouds would pile up as the sun went down, but then the night would be clear and dawn would bring in another perfect day. The barometer remained high and steady; even Colin had rarely known such beautiful sailing.

"Food," I wrote in my journal, "becomes a major interest and it matters far more than usual that it should be good. In fact good fresh food on a plate and plenty to drink are the main things one misses. Also keeping reasonably clean." We still had some apples, potatoes, eggs, and onions, but otherwise most of our fresh food had run out. We carried thirty-five gallons of freshwater and used this only for drinking; our rice or potatoes were boiled in seawater. We all now looked disgusting, with filthy, lank hair and—in Colin's case and mine—increasingly stubbly cheeks.

People sometimes complain of the monotony of the sea, but it is, with the sky, the most changeful of all natural spectacles. Its surface, brushed by the wind, whether gently or with violence,

---

* As well as being a device for measuring speed and distance, the log is—confusingly—also the journal that records a vessel's course, distance run and position (when known), as well as the weather and anything else the watch-keeper feels like mentioning.

presents patterns of infinite variety, and its color too undergoes astonishing transformations, depending on factors like the time of day, the depth of water, and the weather. But despite the ever-changing vistas of sea and sky, time passed very slowly, and I often found the close physical confinement trying. We spent almost all our time either on watch in the narrow cockpit, a space perhaps seven feet by five with the tiller in the middle, or down below in the saloon. This was a bit bigger, but there was very little room to move around, and at more than six feet I could not stand upright. In fine weather I would often sneak off to the foredeck to read. There was nowhere else to go but overboard. We were getting very little physical exercise and this probably contributed to my sense of frustration and impatience: sometimes I felt like screaming.

Colin deliberately did not plot our position on the small-scale North Atlantic chart until we were well on our way, and even then it looked as if we had hardly made any progress. This was not surprising given that we were crossing three thousand miles of ocean at a rate seldom faster than a brisk jog. The possibility that the weather might change for the worse was always on our minds, as we knew we would be lucky not to encounter a gale at some point. The condition of the boat was also a concern—we ran the engine for an hour or so every few days to charge the batteries, pumped out the bilges (counting how many strokes it took to assess how leaky the boat was), and watched the sails and rigging for signs of wear and tear. Sometimes we had to carry out minor repairs. The sliders that attached the mainsail to the track running up the mast often came loose and had to be reattached, and we sewed up a seam on one of the foresails where the stitching had worn out. The steering compass too caused problems: the fluid in which the compass card floated began to leak out and a bubble appeared in the clear plastic bowl that covered it. As the bubble steadily grew it became harder to read the course, so we dismantled the binnacle, found the leak, and patched it up with chewing gum. We then topped up the fluid with gin, but the operation was not a complete success—a small parcel of air obstinately wobbled at the top of the bowl, and the cost in precious liquor was high.

One day a bird landed on *Saecwen*'s deck—some kind of flycatcher, I think—so exhausted that it made no attempt to move when I offered it some water. Eventually it fluttered away. How it had reached us and where it was going was a mystery. As our distance from the land steadily increased, some part of me was always anxiously aware of the immensity of the ocean, the miles of water fading into chilly darkness beneath us, and the almost ludicrous smallness and fragility of our thirty-five-foot boat.

DESPITE OUR COMPASS problems, our sextant and chronometer told us where we were to within a few miles. The exotically named Admiral Sir Cloudesley Shovell was less fortunate when, on the night of October 22, 1707, he entered the English Channel with twenty-one British warships under his command. The fleet drove on to the reef-strewn Isles of Scilly, which were then guarded only by a single lighthouse on the island of St. Agnes, and four ships went down with the loss of some two thousand lives. Shovell himself was washed ashore and reportedly murdered by a local woman who fancied a ring on his finger. This notorious disaster, which has often been cited as evidence of the dangers of navigating without an accurate means of determining longitude, may in fact have been caused by errors in the assessment of the fleet's latitude, or by mistakes in the recorded position of the Scilly Isles[5]—quite possibly both.

Navigation posed many problems in the days before celestial navigation was perfected, and the cause of a wreck can therefore seldom be attributed to any single factor. The notorious loss of the Dutch East India Company ship *Batavia* in 1629 is a case in point.[6] She was wrecked on a reef off the west coast of Australia after crossing the Indian Ocean on her way to Java. The reef in question was part of an extensive group of low islands discovered by a Dutch sailor named Frederik de Houtman in 1619, and the *Batavia* drove on to it under full sail. The lookout had in fact spotted white water ahead but the master, convinced that it was merely a reflection of the light of the moon, refused to alter course or shorten sail. A faulty DR estimate of the *Batavia*'s longitude might well have encouraged this disastrous misjudgment: the master apparently thought he was six hundred miles from land.[7] Such an error would be quite understandable after a long ocean passage, especially since the prevailing westerly winds in the southern Indian Ocean generate an east-going current that would have been hard to detect. To confirm this, however, it would—at the very least—be necessary to know whether the master was using a chart on which the position of the reef was correctly recorded and, if so, where he believed his ship was in relation to it. It is equally difficult to assess the reasons for the loss of the British warship *Ramillies*, which occurred as late as 1760. Having entered the English Channel, she was caught in a gale off the south coast of Devon and embayed. She tried to anchor but was driven ashore, with the loss of all but twenty-seven of her crew of eight hundred. Assuming that her commander knew the latitude correctly, was he mistaken in his longitude? Was the chart he was using in error? Or did poor visibility mean that those on watch failed to notice the approaching coast before it was too late? We have no way of knowing, but the impossibility of accurately determining their longitude put offshore navigators under a heavy handicap.

Commodore George Anson's circumnavigation of the world provides the clearest illustrations of the navigational difficulties that sailors faced before the longitude problem had been solved. England being at war with Spain, Anson (1697–1762) was dispatched in 1740 with a squadron of ships and more than 1,900 sailors and soldiers[8] to harass the enemy's colonial settlements in the Pacific. Long delayed by contrary winds, the squadron reached Madeira—their first port of call—in October, having already been at sea for forty days. The island's longitude was laid down on contemporary charts—as it happens, quite accurately—in 17° West "of London," but Anson and his men placed it somewhere between 18°30' and 19°30'.[9] Their DR was out by at least seventy-five nautical miles. Five months later, in early March 1741, having passed through Le Maire Strait, northeast of Cape Horn, Anson's ships struggled around into the Pacific, facing a succession of terrific gales: men were injured or lost overboard, some lost fingers or toes to frostbite, the ships began to leak heavily, and both sails and rigging were frequently damaged. Typhus and dysentery had already weakened the squadron on the voyage south from Madeira, but scurvy too now began to take its toll.[10] On Anson's flagship, the *Centurion*, the crew were so much weakened that they were unable to throw overboard the bodies of their dead shipmates. One old soldier, who had fought at the Battle of the Boyne in 1690, discovered as he lay dying that his long-healed, fifty-year-old wounds were reopening.[11]

On April 3 the *Centurion* ran into a particularly severe storm:

> *In its first onset we received a furious shock from a sea which broke upon our larboard quarter, where it stove in the quarter gallery, and rushed into the ship like a deluge . . . to ease the stress upon the masts and shrouds we lowered both our main and fore-yards, and furled all our sails, and in this posture we lay to for three days. . . .*[12]

Despite the appalling conditions, Anson and his men were convinced that they had not only rounded Cape Horn but had also made good westerly progress out into the Pacific. Judging that it was now safe to head north, they were in for a nasty surprise. On the night of April 13–14, when they thought they were hundreds of miles out to sea, the leading vessel caught sight of rocky cliffs—probably the western end of Noir Island, off the southwest coast of Tierra del Fuego. She fired a gun and showed lights to warn the ships astern of her of the impending danger:

> *[The land] being but two miles distant, we were all under the most dreadful apprehensions of running on shore; which, had either the wind blown from its usual quarter [south-west] with its wonted vigour, or had not the moon suddenly shone out, not a ship amongst us could possibly have avoided. . . .*[13]

One officer recalled seeing the cliffs rearing up "like two black Towers of an extraordinary height," but every ship managed to get clear.[14]

Anson's own log gave the *Centurion*'s longitude on April 13 as 87°51' W, while another surviving log gives a longitude of 84°12' W just before land was sighted.[15] The difference between these estimates is itself an indication of how difficult it was to determine longitude reliably by DR. In fact the longitude of Noir Island is about 73 degrees West, which means that Anson's estimate was out by nearly 14 degrees—a distance of almost five hundred nautical miles in this latitude. The authorized account of the voyage plausibly places the blame for these very large errors on the unexpected strength of the ocean currents in this locality:

> *It was indeed most wonderful that the currents should have driven us to the eastward with such strength; for the whole squadron esteemed themselves upwards of ten degrees*

> *more westerly than this land, so that in running down, by our account, about nineteen degrees of longitude, we had not really advanced above half that distance.*[16]

The squadron—from which two ships had already separated—faced further battering by storms as it struggled northward. By the end of April, as the death toll from scurvy rapidly mounted, each of the surviving vessels found itself alone. Having barely survived a hurricane at the end of May off the island of Chiloé,[17] the *Centurion* headed for a planned rendezvous at the island of Juan Fernández (now Robinson Crusoe Island) off the coast of Chile, where desperately needed fresh provisions could be found. However, to save vital time, Anson "resolved, if possible, to hit the island upon a meridian [of longitude]."[18] In other words, rather than heading north up the coast of Chile and then running down the island's latitude in the time-honored fashion, he took the chance of sailing directly for it.

On May 28 they had nearly reached the latitude in which the island was laid down and "had great expectations of seeing it: But not finding it in the position in which the charts had taught [them] to expect it" they were afraid that they might have gone too far to the west.[19] Though Anson himself was "strongly persuaded" that he had glimpsed the island, his fellow officers were unconvinced and, following "a consultation," it was agreed that they should head back to the east. They sighted the distant, snowcapped peaks of the Chilean cordillera on May 30:

> *Though by this view of the land we ascertained our position, yet it gave us great uneasiness to find that we had so needlessly altered our course when we were, in all probability, just upon the point of making the island; for the mortality amongst us was now increased in a most dread-*

> *ful degree, and those who remained alive were utterly dispirited by this new disappointment. . . .*[20]

It took nine days to regain the ground they had lost, and it was not until June 10 that the *Centurion*, whose crew at full strength would have numbered between four and five hundred men, at last reached Juan Fernández "with not above ten foremast men in a watch capable of doing duty." This single navigational error had cost the lives of "between seventy and eighty of our men, whom we should doubtless have saved had we made the island [on May 28], which, had we kept on our course for a few hours longer, we could not have failed to have done."[21]

Once the remnants of his squadron had gathered at Juan Fernández and the surviving members of the ships' crews had recovered their strength, Anson carried out a raid on a small Spanish coastal settlement in Peru and also captured a few merchant vessels. He next tried to intercept a Spanish treasure ship that was expected to sail from Acapulco, but news of his presence had reached the Spanish authorities, and the ship remained safely in port. Disappointed but undaunted, in May 1742 he set out across the Pacific with his two surviving ships.[22] In the course of an agonizingly slow voyage, again greatly complicated by unreliable charts and uncertainties about their longitude, one ship had to be abandoned, and the *Centurion*, with eight or ten men dying every day "like rotten sheep," was barely afloat when she reached Tinian in the Mariana Islands in August.[23]

Anson's determination now, at last, paid off. He was able to reach Canton (modern Guangzhou), where the *Centurion* was repaired, and then succeeded in capturing the treasure ship *Nuestra Señora de Covadonga*, off the Philippines. The *Covadonga* was carrying more than 1.3 million pieces of eight and 35,682 ounces of silver. The exact value of the loot Anson finally brought home in 1744 is uncertain, but as a ship's captain as well as commander-in-chief, he would have received three-eighths. His share of the treasure from the *Covadonga* alone may have amounted to the vast sum of £91,000. For comparison, the pay due to him in the course of the whole voyage fell just short of £720.[24] The captured treasure was paraded through the streets of London to national jubilation, and little attention was paid to the fact that 1,400 men had failed to return home. Propelled by this success, Anson was later ennobled and rose to the very pinnacle of the Royal Navy, serving twice as First Lord of the Admiralty.

Chapter 6

The Marine Chronometer

Day 7: Stayed in my bunk 'til 0745 and then sat in the sun reading Slocum while watching clouds building up—the barometer is falling and the weather is starting to break. The wind veered to W and increased to F 5 so we took down the main and ran on under genoa at 6–7 knots. Colin was still not comfortable, though, so we went right down to the pocket handkerchief No. 2 stays'l, which cut our speed to 4 knots.

After lunch—the usual sandwiches though the bread is getting moldy round the edges—I went to sleep or tried to for two hours. Much rolling and rattling. It's amazing how much the weather affects one's mood out here. All the same, we've made good progress so far and today we've covered 144 miles, noon to noon. Our course is now 105° magnetic. Helped Colin work out our noon position: 42°34' N, 46°16' W.

Celestial navigation would be easy if the sun and all the other heavenly bodies stood motionless in the sky—as Polaris does, almost.* It would then be possible to fix your position by sextant sights without the need to know the time or even the date. The cosmos, however, is not that obliging. Not only does the earth rotate completely once every day, but its axis of rotation—currently inclined at roughly 23.5 degrees to the plane of its or-

* In fact it is not quite motionless: it revolves around the celestial north pole in a tight circle.

bit around the sun—also changes gradually over a cycle of about 25,800 years.* To complicate matters further, the earth's orbit around the sun is elliptical rather than circular, with the result that the interval between one passage of the sun over the observer's meridian and the next is not quite constant. So not only do the heavens appear to be in motion, but that motion itself is also changeful. This is most obviously revealed by the variations in the path of the sun across the sky—which is measured by its declination to the north or south of the equator—the phenomenon that gives rise to the seasons. The behavior of the planets and the moon is yet more complex.

The ancient Greeks and Romans, who leaned heavily on earlier Babylonian learning, had a well-developed understanding of the paths that the various heavenly bodies described, as did the Arab astronomers who followed them. They clung, however, to the view—associated with the astronomer Ptolemy (*c.*90–168 CE)—that the earth was at the center of the universe, and this theory prevailed until the time of Copernicus (1473–1543).[1] Though Ptolemaic orthodoxy may have been misguided, it did not prevent astronomers producing accurate solar declination tables as far back as the end of the fifteenth century. These enabled mariners for the first time to adjust their observations of the sun to allow for the seasonal changes in its meridian altitude. Now they could determine their latitude at noon as the sun crossed their meridian, as well as after dark (from the height of Polaris), subject to the limitations of the instruments then at their disposal. Moreover, they could continue to find their latitude when south of the equator—when Polaris had disappeared below the northern horizon. This breakthrough helped the Portuguese to open up an enormously valuable trade route into

---

* A process that gives rise to the "precession of the equinoxes." It also wobbles slightly, an effect known as "nutation."

the Indian Ocean around the Cape of Good Hope. Early in the sixteenth century the Portuguese also devised a rule for determining latitude by reference to the stars of the Southern Cross—which lie some distance from the south celestial pole.[2]

While latitude could be determined quite easily, the earth's motions meant that the measurement of longitude was a much more difficult challenge. Early in the sixteenth century, the astronomer Gemma Frisius (1508–55) realized that a promising approach to solving the longitude problem would be to find a way of measuring time accurately—whether on land or sea. An observer equipped with an accurate enough clock set to the time at a reference meridian could in principle compare the time of an event (such as sunrise or sunset) with the *predicted* time of the same event at a reference meridian—such as Greenwich (the internationally agreed "prime meridian" since 1884). The observer's longitude could then be derived by converting the *time difference* in hours and minutes into a *spatial displacement* measured in degrees and minutes east or west—one hour being equal to 15 degrees of longitude (360 divided by 24).

It was not until the early seventeenth century that Copernican theory was firmly established on the basis of the observations of Tycho Brahe (1546–1601), Galileo Galilei (1564–1642), and Johannes Kepler (1571–1630). Galileo's momentous discovery of the moons of Jupiter and, soon afterward, of the changing phases of the planet Venus, not only provided overwhelming evidence that the earth was not the center of the universe but also opened the way to a proper understanding of planetary motion.[3] The invention of the first pendulum clock in the 1650s by Christiaan Huygens (1629–95) also marked a great advance. It was now possible for astronomers to measure time with sufficient precision to predict with great accuracy the positions of all the major heavenly bodies day by day—though, as we shall see, there was one troublesome exception: the moon. The establishment of the two great royal observatories in Paris (1667) and Greenwich (1675) contributed notably to this process. These technical developments, coupled with major theoretical advances—of which the publication in 1687 of Newton's laws of motion was the most significant—were crucial steps on the path to the eventual solution of the longitude problem.

By the end of the seventeenth century, the laborious observations of astronomers had yielded the first accurate ephemeris tables.[4] An observer on dry land supplied with a pendulum clock could now regulate it by reference to the predicted events and thereby establish his or her longitude. French scientists were the first to apply the new technology to the making of accurate terrestrial maps, and the results were sometimes surprising. In 1693 a new map of the coast of France based on an elaborate survey supervised by the astronomers Jean Picard (1620–82) and Philippe de La Hire (1640–1718) revealed that the kingdom had shrunk. The port of Brest, for example, had moved fifty miles to the east of its position on the best existing map. King Louis XIV is reputed to have complained that he had lost more territory to his astronomers than to his enemies.[5]

The French undertook much basic research—including heroic efforts to determine the precise shape of the planet, a knowledge of which was essential if maps were accurately to reflect reality. Scientists were sent all over the world in an attempt to decide whether Newton's prediction that the earth bulged slightly around the equator was correct. If it did, the geographical length of a degree of latitude would increase as one moved away from the equator toward the poles. While one such expedition went to Finland and another to South Africa, a third, led by Louis Godin, set out in 1735 for the Andes to try to measure a degree of latitude at the equator. Godin and his team endured extraordinary hardships, first struggling through tropical jungles and then working at heights of over sixteen thousand feet on the freezing mountaintops, as they carefully measured baselines and extended a network of triangles along the mountain chain over a distance of some two hundred miles.* Their efforts, combined with the work of the other expeditions, confirmed Newton's prediction.[6]

The British were initially slow to learn from the French mapmakers, but a growing awareness of the great military and commercial advantages conferred by good maps and charts prompted action. Murdoch Mackenzie Sr. (1712–97) led the way with his pioneering marine survey of the Orkney Islands in the 1740s, based on a rigid system of triangulation using precisely measured baselines, the first of which was laid out on the frozen surface of a lake.[7] Mackenzie's were the first accurate British charts, and he also invented the system of symbolic abbreviations that survives to this day. His *Treatise on Maritim Surveying* (published in 1774) was to set the pattern for every hydrographic survey conducted over the next hundred years, and in it he listed the quadrant and sextant as essential items of the marine surveyor's equipment. Of the sextant he had this to say:

> *This instrument may be used with great Advantage in Maritim surveys, on most Occasions; being more portable, more readily applied to the taking of Angles, and generally more accurately and minutely divided than Theodolites are: an Observer is less liable to make mistakes with it; and, which is a very material advantage, he can take angles with it at sea, as well as on Land.*[8]

---

* They did not complete their work until 1743, and one of their number, Charles-Marie de La Condamine, daringly returned home by crossing the Andes and descending the Amazon to the ocean on a raft—the first European ever to make this journey. There was romance, too: Godin fell in love with the thirteen-year-old daughter of the Spanish viceroy of Peru, and they eventually married after a twenty-year separation. See Danson 37–42.

Mackenzie is also credited with the invention of the "station pointer"—an invaluable instrument that enables the coastal navigator quickly to fix his position by taking horizontal sextant angles between three or more fixed points.

Not until the 1790s did the newly established Ordnance Survey follow Mackenzie's example and begin mapping Britain by triangulation. In 1797 the intricate network of triangles was extended from Land's End to the Scilly Isles and, to general consternation, it emerged that the position of the islands shown on contemporary charts was out by the astonishingly wide margin of twenty nautical miles.[9] Cold comfort for poor Shovell.*

FOR ALL THE progress that was being made in mapping the land, accurate position-fixing at sea still remained an impossible dream in the early eighteenth century. In fact it was no closer to reality than it had been 150 years earlier when the Spanish, conscious of the vital commercial importance of their overseas colonies and the difficulties surrounding accurate and therefore safe navigation, began to seek a shipboard solution to the longitude problem. In 1567 King Philip II offered the first cash prize to anyone who could crack it, and in 1598 his successor, Philip III, raised the stakes: the winner would receive a one-off payment of 6,000 ducats together with an annual pension of 2,000 ducats. So important was the goal that this princely annuity was promised even to the *heirs* of the eventual winner.[10] Other prizes were later announced by the Dutch and Venetian republics, by France, and, eventually, by Britain. Under the terms of the British Longitude Act of 1714, a sum of up to £20,000 was offered

---

* It is a curious fact that the position of the island of Tahiti—on the far side of the world and discovered by Captain Samuel Wallis only in 1767—was by then much more accurately known.

as "a due and sufficient Encouragement to any such Person or Persons as shall discover a proper Method of Finding the said Longitude." This would now be worth several million pounds.

The Longitude Act, however, imposed high standards of accuracy: to win the maximum amount the successful method had to be capable of determining longitude within a margin of error not exceeding half a degree of a great circle (equivalent to thirty nautical miles). Half the maximum prize would be payable when the commissioners of the new Board of Longitude were satisfied that the proposed method extended to "the Security of Ships within Eighty Geographical Miles from the Shores, which are Places of the greatest Danger," while the balance would be paid "when a ship . . . shall actually Sail over the Ocean, from Great Britain to any such Port in the West-Indies, as those Commissioners . . . shall Choose or Nominate for the Experiment, without Losing their Longitude beyond the Limits before mentioned." Moreover, the reward would be paid only "as soon as such method for the Discovery of the said Longitude shall have been Tried and found Practicable and Useful at Sea, within any of the degrees aforesaid." The words "Practicable" and "Useful" were to give rise to bitter disputes. Lesser rewards were available for proposals that the Commissioners judged of "considerable Use to the Publick." [11]

Though pendulum clocks coupled with the new ephemeris tables permitted land-based observers to determine their longitude accurately, nobody had yet managed to devise a time-keeper that could be relied on at sea. Existing spring-driven clocks and watches were hopelessly erratic, and despite valiant attempts, it proved impossible to make pendulum clocks work reliably on board ship. Strenuous efforts were made to find methods of determining longitude that did not rely on astronomical observations and could therefore be employed without the need to know the time. Mapping the geographical variations in the direction of the earth's magnetic field seemed to offer some hope, but in the end this line of inquiry proved abortive and the heavens became the exclusive focus of scientific attention among those seeking to solve the longitude problem. If seagoing clocks were not to be relied on, then perhaps the sailor could find the time from observations of the sun, moon, and stars. The challenge was to identify a frequently occurring astronomical event the precise time of which could be both accurately predicted *and* easily observed on board ship, anywhere in the world. Published tables of the predicted times of such events would in principle enable the navigator to find the time at a given reference meridian (such as Greenwich or Paris) wherever he happened to be—providing the skies were clear. Comparison with the local time—derived from astronomical observations—would then reveal the observer's longitude.

Various methods of achieving this goal had already been suggested. For example, in 1616 Galileo opened discussions with Spanish officials about the possibility of using observations of the appearance and disappearance of the moons of Jupiter (the four largest of which he had recently discovered) as a means of determining the time at the reference meridian. In return for a large fee for travelling to Spain to demonstrate his method to King Philip III, an annual royalty both for himself and his heirs, as well as appointment to the chivalric Order of Santiago, he proposed to draw up the necessary tables and update them annually; he even invented a telescopic device to be worn on the head that was supposed to permit making the necessary observations at sea.[12] But the Spanish lost interest, and in 1635 an aging Galileo turned to the Dutch, this time with improved tables and a mechanical device for representing the motions of the Jovian satellites that he called the "Jovilab."

The Dutch States General responded enthusiastically and even appointed an astronomer to act as a technical go-between, but Galileo—who was by now going blind—was unable to generate the orbital parameters of the moons on which sufficiently precise predictions could be based.[13] In any case, a fairly powerful telescope was required to observe the moons of Jupiter, and such an instrument could not be held steadily enough on board ship. And there was another problem lurking in the background: without an accurate shipboard timekeeper, how exactly was the navigator supposed to compare local time (derived most easily from sun sights) with the time obtained from the tiny Jovian moons—visible only after the sun had set? Galileo claimed he knew how to make a sufficiently accurate pendulum clock, but he had not succeeded in doing so by the time he died, and anyway it would have been of no use at sea.[14]

Jupiter's moons were, however, very useful to land-based observers equipped with pendulum clocks—once the necessary tables had been produced at the Paris Observatory. In the 1680s a French expedition established the longitude of the Cape Verde Islands, Guadeloupe, and Martinique using this technique,[15] and Picard and La Hire also employed it when making their map of France. Eclipses of the sun were among the other possibilities, but they were too infrequent to be of much use, and it was not until the invention of the sextant that it was possible to observe them with sufficient accuracy on board ship. Spanish navigators and astronomers had experimented with the technique in the sixteenth century, but the results, even at land-based observatories, were of no value.*

So it is not surprising that when the Longitude Act was passed in 1714, few observers expected that anyone would soon succeed in claiming the big money. Many bizarre and frankly ludicrous

* Occultations of stars by the moon could be observed with greater ease, but—as we shall see—the necessary predictions of the moon's movements were not yet available.

proposals were put forward, and in consequence the quest became something of a standing joke. William Hogarth included a cheerful lunatic searching for a solution to the longitude problem in the background of the scene from the madhouse in *The Rake's Progress* of 1735.

Such skepticism was misplaced. After a struggle lasting hundreds of years, two radically different solutions to the problem of finding the time on board ship emerged almost simultaneously in the 1750s—one mechanical, the other astronomical. Both, however, relied on accurate angular measurements made with a quadrant or, better still, a sextant. As we shall see, one method was based on a new kind of clock, while the other depended on the first accurate tables of the motions of the moon. In practice, however, the two techniques were to be mutually dependent for many years to come.

THE EXTRAORDINARY STORY of the development by John Harrison (1693–1776) of the first accurate shipboard timekeeper—and his long struggle for official recognition of his feat—is by now well-known. In 1759, after more than thirty years of experimentation, he produced a highly innovative "watch" (known as "H4"). It was not regulated by a pendulum and exploited the ingenious principles of compensation he had developed in earlier experimental devices; it was also a great deal smaller and more practical. H4 easily passed the second (and possibly also the first)[16] of two rigorous sea trials—a voyage to Barbados and back in 1763. But the Longitude Board was cautious about recognizing Harrison's remarkable technical breakthrough. Rigorously interpreting the wording of the Longitude Act, they demanded to be convinced that the new watch's exceptional performance had been more than a fluke and that the mechanism itself could be reliably replicated at an affordable price. Arguments about whether H4 and its maker had or had not satisfied the precise terms of the act were to drag on for years. The elderly Harrison, vigorously supported by his son, William, was enraged by the apparently perverse delays in awarding him the full prize of £20,000, and by adjustments to the terms of the act that—in his view—seemed designed to deny it to him. His tactless and explosively ill-tempered behavior alienated many members of the Board, which had by the end of 1762 already funded his labors to the tune of £4,750—a very substantial sum.

As guardians of public funds, the board members were understandably anxious not to expose themselves to charges of waste. But their reluctance to reward Harrison in full was also influenced by the belief—which Newton had shared—that the only reliable solution to the longitude problem must be astronomical, not mechanical. After all, watches could go wrong, and seemed very likely to do so in a damp and bumpy ship at sea—especially if the temperature varied a good deal, as it would on a voyage from Europe to the tropics. How would the navigator be able to tell if the watch started to misbehave? Who was going to fix it if it stopped? How could any errors be corrected?

These were perfectly fair questions, and as experience subsequently showed, many chronometers did indeed perform poorly, often running irregularly or stopping altogether for no obvious reason. Even their own makers did not understand exactly what they were doing: they were artists as much as engineers, and they relied heavily on trial and error. The sun, moon, and stars, by contrast, were the very embodiment of perfection—and indeed the basis of time itself. Until the invention of the "atomic clock" in the mid-twentieth century, the movements of the sun and stars remained the fundamental indices of time. Harrison's watch may well have seemed inelegant to the more mathematically minded members of the Longitude Board—a questionable, brute-force solution to a problem they regarded as essentially astronomical in nature.

Harrison's great achievement was to demonstrate that it was indeed possible to build a watch that could do the job—not just in stable conditions on land, but also in the wildly variable environment of a ship at sea. He was certainly a difficult and irascible man, as was his son, but he was highly ingenious and extremely determined, and in 1773, following a powerful speech in the House of Commons by Edmund Burke, and a sympathetic intervention by King George III himself, Parliament (rather than the unbending Longitude Board) awarded him a further £8,750.[17]* The practical marine chronometers (as these "timekeepers" were eventually to be known) that his pioneering work inspired were, however, very different from H4. They owed much to the inventive skills of other watchmakers like Pierre Le Roy and Ferdinand Berthoud in Paris and Larcum Kendall, John Arnold, and Thomas Earnshaw in London.[18]

The chronometer we carried aboard *Saecwen* was a descendant of those developed in the last decades of the eighteenth century and probably differed little from them. It sat luxuriously in a pretty mahogany box with brassbound corners, secured by strong elastic cords in a safe corner of the cabin near the mast. Lifting the lid, a circular brass case was revealed, with a plain but elegant dial and thin, spear-shaped hands, the whole mechanism supported in a gimballed cradle that isolated it quite effectively from the motion of the boat. Colin alone undertook the delicate task of winding it, a ritual performed at the same time each day in order to maintain an even tension in the main-

---

* This meant that Harrison had in the end received rather more than £20,000 from the public purse, though he had incurred significant expenses and complained that he was still £1,250 short.

The "PZX" Triangle

The "PZX" triangle is at the heart of celestial navigation and can be used to solve a variety of navigational problems.

The angle XPZ is the key to finding the "local time at ship." P is the North Pole, *X* the "geographical position" of the sun, and Z the position of the ship. The arc XA is the declination of the sun (tabulated in the *Nautical Almanac*); the arc ZB is the ship's latitude (typically obtained from a "mer alt"). We can calculate the lengths of sides PX and PZ: PX is 90 degrees minus the sun's declination, while PZ is 90 degrees minus the ship's latitude. The third side, ZX, is equivalent to the "zenith distance" of the sun, which is obtained by subtracting its altitude (observed with the sextant) from 90 degrees.

Using spherical trigonometry we can now derive the angle XPZ, which is the sun's Local Hour Angle or LHA – in this case a measure of the time elapsed since the sun crossed the ship's meridian. The time that has passed since the sun crossed the Greenwich Meridian (revealed by the chronometer) is its Greenwich Hour Angle or GHA. By subtracting the LHA from the GHA the navigator can obtain the required 'local time' and thereby the ship's longitude. Similar calculations can be performed using other celestial bodies.

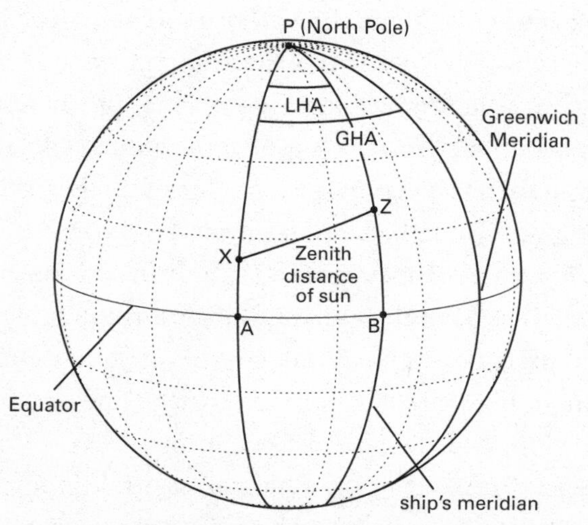

spring. The chronometer's lovely, silky tick was like a breathless heartbeat. With the sextant, it was a thing of beauty.

USING OUR CHRONOMETER (duly set to Greenwich time) I learned from Colin a rough-and-ready method of determining *Saecwen*'s longitude. When the weather was clear I would time the moment of sunrise (or sunset) and compare the results with the times of these events tabulated in the *Nautical Almanac*. If, for example, the disc of the sun appeared over the eastern horizon at 0400 GMT according to the chronometer, while the tabulated time of the same event at Greenwich was 0600, it followed that we were two hours—or 30 degrees—west of Greenwich. The results I obtained were—at best—accurate to about half a degree either way. In principle the same technique could be used to obtain the longitude by comparing the local and Greenwich times of a heavenly body's transit across our meridian.

In practice, however, it is difficult to determine the *exact* moment of sunrise because atmospheric refraction, which is strongest at low angles, has the effect of "lifting" the sun's disc so that it remains visible for some time after it has actually dipped below the horizon. The timing of a meridian passage at sea is also problematical as heavenly bodies pause for a significant interval at the height of their arc. One way of doing so is to take two timed "equal altitude" observations of the relevant body on either side of the meridian and to halve the time difference between them. A major drawback of this method is that clouds may obscure the crucial second sight. It also requires a reasonably accurate clock. And precision is vital: an error of just one minute in the measurement of either local or Greenwich time can result in a positional error of as much as fifteen miles.

How then is the navigator to obtain an accurate longitude, even if he or she has an accurate clock? Knowing the exact time at the reference meridian by itself is no help. There has to be something with which that time can be compared. The solution lay in discovering the *local time* at ship from sextant observations—usually altitudes of the sun in the morning or afternoon, though other heavenly bodies could be used. Mathematicians developed a variety of techniques for achieving this objective, all of which involved solving what came to be known as the "PZX" triangle—see diagram, page 69. These methods—which, as we shall see, relied on knowing the ship's latitude—remained at the heart of celestial navigation until the emergence of the "new navigation" in the 1870s.

Chapter 7

Celestial Timekeeping

Day 8: Up again at 0400 and got a sunrise longitude fix of about 45° W at 0515. The same weather—force 5 from WSW with a fair bit of sunshine interspersed with low cloud and rain showers. Much rolling and rattling of crockery and cutlery. Not much speed—only 4 knots.

One week at sea. I tried to measure how far we had gone but failed to realize that on the small-scale North Atlantic chart the latitude scale is not uniform, so I got it wildly wrong. Colin filled in the track chart. We have done 830-odd miles but there's a long way to go.

The change of weather and the prospect of at least two more weeks at sea is depressing. At noon our position was 42°30' N, 43°57' W, making a day's run of 102 miles. Not all that fast.

While Harrison labored, mathematicians and astronomers across Europe were also trying hard to develop a method of determining the time by the "lunar-distance method." The theory underlying the use of "lunars" was that the angular distance between the moon and the sun (planets and certain stars could also be used, with some loss of accuracy) changed so rapidly and predictably that it could be used like a celestial clock—a clock that told the same time anywhere in the world, though it was, of course, not always visible. But the complicated behavior of the moon—powerfully influenced as it is by the gravity both of the sun and of the earth—made it much harder to predict its celestial coordinates with accuracy than those of the other heavenly bodies.

Although Newton had dazzled the world with the laws of motion that allowed the paths of the sun and its planets to be predicted with hitherto unimaginable precision, the moon had defeated him. But though his lunar tables were not good enough for the purposes of determining longitude, Edmond Halley (1656–1742) recognized that the errors in them recurred regularly every eighteen years and eleven days—in accordance with a well-known cycle of eclipses. This discovery enabled him to develop a rule for correcting the tables, which was later improved by the French astronomer Pierre-Charles Le Monnier (1715–99). There were still imperfections, but in 1750 another Frenchman, Alexis Claude de Clairaut (1713–65), published a new theory of the moon's motion in response to a competition launched by the St. Petersburg Academy in Russia. Though the predictions that accompanied it were incomplete, this represented a major step forward. The great Swiss mathematician Leonhard Euler (1707–83), building on Clairaut's work, published his own theory of lunar motion in 1753, and it was this that enabled a young, self-taught German astronomer named Tobias Mayer (1723–62) to achieve a major breakthrough.[1] The effort to find an astronomical solution to the longitude problem was thus a model of scientific internationalism.

Appointed professor of mathematics at the University of Göttingen in 1751, Mayer closely analyzed the available observational data and, with the help of Euler, prepared new tables of the moon's motions that proved far more accurate than any previously available. His work first was already under discussion in England as early as 1754. In 1755 the Astronomer Royal, Dr. James Bradley, reported to the board on his examination of Mayer's tables:

> *In more than 230 comparisons which I have already made I did not find any difference so great as 1½' between the observed longitude of the Moon and that which I computed by the tables . . . it seems probable that, during this interval of time, the tables generally gave the Moon's place true within one minute of a degree.*
>
> *A more general comparison may perhaps discover larger errors; but those which I have hitherto met with being so small, that even the biggest could occasion an error of but little more than half a degree of longitude, it may be hoped that the tables of the Moon's motions are exact enough for the purpose of finding at sea the longitude of a ship, provided that the observations that are necessary to be made on shipboard can be taken with sufficient exactness.*[2]

With Mayer's tables it seemed possible that a practical, astronomical method of determining the time accurately on board ship might at last be within reach. To make this lunar-distance method work successfully, however, a very accurate device for measuring angular distances was essential. The standard Hadley quadrant could measure angles only up to ninety degrees, and since the angular separation of the sun and moon often exceeded that amount, a larger instrument would plainly be helpful. Mayer himself had proposed the use of a circular device—the "reflecting circle"—but when Captain John Campbell of the Royal Navy was testing Mayer's tables at sea in 1757 he found it inconvenient to use. Campbell then came up with the simple idea of enlarging Hadley's quadrant, and so commissioned a leading instrument maker, John Bird, to make the very first sextant.[3] Bird's device was not only bigger but more accurate than the standard quadrants of the day: its frame was made of brass rather than wood, and it was equipped with a telescope instead of simple aperture sights. While the quadrant was, in general, capable only of measuring to the nearest minute of arc, the sextant, with the help of a tangent screw and vernier, could in principle give readings accurate to 10".[4] Using the new instrument at sea off Ushant in 1758 and 1759, Campbell found it was possible to obtain lunar distances on fifteen to sixteen days a month rather than just eight with a quadrant, and more important still, his observations with the new instrument yielded longitudes accurate to within 37 minutes—well within the terms of the Longitude Act's provisions.[5] With the development of the mechanical "dividing engine," perfected by Jesse Ramsden in 1775, it became possible to graduate the scales of instruments like the sextant with much greater precision than was possible by hand. Not only was accuracy improved, but manufacturing costs were also much reduced. The Longitude Board accordingly rewarded Ramsden's work with prizes amounting to more than £1,000.[6]*

The astronomer Nevil Maskelyne (1732–1811) was a central figure in the struggle to solve the longitude problem, and a controversial one. He sailed to St. Helena in 1761 in the hope of observing the Transit of Venus as part of an international effort to exploit this rare event, which occurs (twice, eight years apart) at intervals of more than one hundred years. The aim was to make simultaneous observations from widely separated parts of the world and to use the results to determine the distance of the earth from the sun. Poor weather frustratingly prevented him from making the crucial observation of the planet crossing the sun's disc, but the two long voyages nevertheless proved useful. They gave him ample opportunity to familiarize himself with the general challenges of shipboard navigation, and also enabled him to test the lunar-distance method at sea us-

* Improvements to the quality of mirrors, shades, and lenses also added to the accuracy of sextants during the latter part of the eighteenth century.

ing Mayer's as yet unpublished tables. He took at least sixteen lunar-distance observations on the outward voyage in an East India Company vessel, and although bad weather prevented him taking any during the last eight days, he reported that his longitude error on arrival at St. Helena was only 1½ degrees—far less than that of some of the ship's officers, who were presumably relying on DR.

Maskelyne reported to the Royal Society that "a person who will take the pains necessary to make accurate observations and has at the same time leisure and ability to make the requisite calculations will be able to ascertain his Longitude by this method as near as will in general be required."* On the return voyage to England, Maskelyne successfully encouraged some of the ship's officers to learn and practice the lunar-distance method. In the light of this experience he concluded that longitude could always be found using the lunar method to "within a degree, or very little more": he was to remain one of its most devoted champions.[7] In 1763 he published the *British Mariner's Guide*, which explained how to make lunar-distance observations and deduce the longitude from them, as well as containing the first English edition of Mayer's tables.

But observing lunars was, initially at least, a laborious and time-consuming process. Four people were ideally required—two to take simultaneous altitudes of the observed bodies, one to record the time intervals between the observations (for which an ordinary watch was sufficient), and one to observe the lunar distance itself—though an expert could, if necessary, manage on his own.[8] Large and variable allowances had to be made for the effects of atmospheric refraction and parallax. This com-

---

* Maskelyne himself fixed the longitude of St. Helena during his stay there and presumably estimated the longitude error on his arrival there retrospectively. I am grateful to members of the Cambridge Digital Longitude Project for advice on this point.

plex process was known as "clearing the distance," and, over the years, at least forty different methods were proposed, though probably few of them were ever used.[9] Having accurately measured the lunar distance, and "cleared" it, the navigator could use Mayer's tables to determine the time at Greenwich. Coupled with observations for local time, the observer's longitude could then readily be established. But the lunar-distance calculations were lengthy and, until precomputed values were published in the new *Nautical Almanac* (discussed on page 80), might take several hours to complete.

Mayer's tables were in due course significantly refined,[10] and it eventually proved possible for a skillful observer using a good sextant to find the Greenwich time using lunars to within one or two minutes—a far narrower margin than that promised by Maskelyne. Such levels of accuracy were adequate for most navigational purposes, but for the making of charts a higher degree of precision was desirable. By taking a long series of observations from temporary observatories set up onshore and averaging the results, the effects of observational errors could be greatly reduced. Navigators on voyages of discovery would apply this technique whenever possible and might perform a hundred or more sets of lunar observations in order to establish accurate fixed points on which to base their charts. Observations of Jupiter's moons as well as eclipses of the sun and moon were also employed when opportunities offered.

MASKELYNE HAS SOMETIMES been painted as the villain who denied Harrison his just deserts, but while his influence over the board (of which he became an ex officio member following his appointment as Astronomer Royal in 1765) was no doubt powerful, the decision to withhold the full reward certainly did not rest with him alone. Relations between Harrison and the

Longitude Board began to deteriorate after the first trials in Jamaica, but soon Maskelyne became the focus of his anger and suspicion. On his arrival in Barbados, Harrison's son William, representing his elderly father, who had stayed at home in England, asserted that the astronomer was not a proper person to conduct the observations necessary to test H4's performance. He claimed that Maskelyne, as a well-known advocate of the rival lunar-distance method, was himself a candidate for the Longitude prize and therefore an interested party.[11]

Maskelyne was, it is true, a devotee of the lunar method, but he did not stand to gain financially from the experiment since it was Mayer's method he was testing, not his own. In any case, there is no evidence that Maskelyne ever took advantage of his official position to enrich himself. Nor did he allow his own views to color his judgment of H4's performance: he later acknowledged in an unpublished autobiographical note that "Harrison's watch was found to give the Longitude of the island [Barbados] with great exactness."[12] That he was among those who proposed William Harrison for the great honor of fellowship of the Royal Society shortly after their return from Barbados[13] does not suggest that he was hostile toward either him or his father, even if his attitude tended to be condescending.

However, in 1767, following extensive tests conducted at Greenwich, Maskelyne reported to the Longitude Board on the performance of H4 in terms that infuriated Harrison:

> *. . . Mr. Harrison's Watch cannot be depended upon to keep the Longitude within a Degree in a West India Voyage of six weeks; nor to keep the Longitude within half a degree for more than a few days; and perhaps not so long, if the Cold be very intense; nevertheless . . . it is a useful and valuable invention, and, in conjunction with observations*

> *of the Distance of the Moon from the Sun and fixed Stars, may be of considerable advantage to Navigation.*[14]

Harrison dismissed Maskelyne's findings out of hand and did not hesitate to publish his complaints. Though Harrison's criticisms of the trials are largely unconvincing, it is fair to say that Maskelyne's own conduct was stiff-necked. While the report Maskelyne produced for the board was perfectly accurate in a technical sense, he made no allowance for the fact that H4 was gaining at a fairly steady rate, at least at the start of the trials. Had he done so, H4 would have passed four out of six of them. Harrison, however, had failed to draw Maskelyne's attention to this problem, and his sulky and uncooperative behavior did nothing to help his case.[15] Nevertheless, even if Maskelyne questioned the reliability of the new watch, it is clear from his testimony to the Longitude Board that he recognized its potential value.

The Longitude Board may perhaps have treated Harrison ungenerously, but this does not mean—as Harrison himself passionately believed—that they were biased in favor of astronomical methods. Having made such a large investment in Harrison's work, they had no reason to turn their backs on it, and their assessment of Mayer's achievements was, in fact, even harsher. Though the efficacy of his tables had been clearly demonstrated, they did not reward him immediately. Interpreting the terms of the Longitude Act at least as strictly as they had in Harrison's case, the board argued that the lunar method was too complicated to be of practical use at sea. In 1765 they declared—with nice symmetry—that while Harrison's chronometer method was "practicable but not generally useful," Mayer's tables were "useful but not practicable." Harrison was therefore deemed to merit an amount not exceeding half the *maximum* award while Mayer's widow (he had died in 1762 at the age of thirty-nine) was entitled to at most half the *minimum* award. She eventually received £3,000 while a mere £300 went to Euler.[16] Harrison, who got £10,000, did very well by comparison.

ONE OF THE greatest achievements of Maskelyne's long and distinguished career, and the one for which he is best remembered, was the development of the *Nautical Almanac*. This was first published—under his personal supervision—in 1766, though it went on sale only in January 1767. Produced continuously ever since, its popularity among mariners all over the world helps to explain why Greenwich was ultimately selected as the prime meridian. Maskelyne subcontracted the tedious and repetitive calculations needed to produce the *Nautical Almanac* to numerous human "computers" around the country, working in their own homes. The most important calculations were duplicated by separate "computers," and then checked by an independent "comparer."

The purpose of the *Almanac* was—quite simply—to give the navigator all the information necessary to find his position at sea, with or without a chronometer. In addition to the standard data that were already available in the French *Connoissance des Temps*, Maskelyne included precomputed lunar distances from the sun at three-hourly intervals (when the two bodies were at the right distance from each other), based on the Greenwich meridian.[17] On the days when the angle between sun and moon was either too great or too small, lunar distances to selected stars were tabulated. These new tables greatly reduced the burden of calculating lunar distances, which might now take as little as half an hour. Maskelyne also published separately the *Requisite Tables*, which included all the information that the navigator needed to clear the distance, as well as a variety of other useful data that did not have to be revised annually.

		JANUARY 1767.			[9]
	Diſtances of ☽'s Center from Stars, and from ☉ eaſt of her.				
Days.	Stars Names.	Noon.	3 Hours.	6 Hours.	9 Hours.
		° ′ ″	° ′ ″	° ′ ″	° ′ ″
1					
2	α Pegaſi.	46. 41. 15	44. 57. 51	43. 14. 53	41. 32. 32
3		33. 15. 35	31. 40. 16	30. 6. 42	28. 35. 2
4	α Arietis.	57. 55. 16	56. 6. 21	54. 17. 44	52. 29. 25
5		43. 32. 47	41. 46. 31	40. 0. 36	38. 15. 1
6	Aldebaran.	62. 4. 49	60. 22. 21	58. 40. 17	56. 58. 37
7		48. 36. 32	46. 57. 27	45. 18. 47	43. 40. 35
8		35. 37. 28	34. 2. 38	32. 28. 29	30. 55. 5
9		23. 22. 20	21. 55. 18	20. 30. 0	19. 7. 3
10	Pollux.	51. 3. 14	49. 27. 59	47. 52. 57	46. 18. 9
11		38. 27. 43	36. 54. 20	35. 21. 12	33. 48. 17
12	Regulus.	62. 42. 22	61. 9. 30	59. 36. 47	58. 4. 13
13		50. 23. 35	48. 51. 53	47. 20. 18	45. 48. 52
14		38. 13. 40	36. 43. 0	35. 12. 28	33. 42. 3
15		26. 11. 51	24. 42. 9	23. 12. 34	21. 43. 10
16	Spica ♍	68. 17. 41	66. 48. 34	65. 19. 30	63. 50. 31
17		56. 26. 28	54. 57. 51	53. 29. 15	52. 0. 41
18		44. 38. 16	43. 9. 50	41. 41. 25	40. 13. 0
19		32. 50. 51	31. 22. 21	29. 53. 51	28. 25. 19
20		21. 2. 16	19. 33. 33	18. 4. 47	16. 36. 0
21	Antares.	54. 40. 6	53. 9. 18	51. 38. 17	50. 7. 5
22		42. 27. 36	40. 54. 57	39. 22. 2	37. 48. 50
20	The Sun.	120. 36. 39	119. 14. 38	117. 52. 30	116. 30. 15
21		109. 36. 50	108. 13. 39	106. 50. 14	105. 26. 38
22		98. 25. 11	97. 0. 7	95. 34. 48	94. 9. 12
23		86. 56. 45	85. 29. 15	84. 1. 25	82. 33. 14
24		75. 6. 56	73. 36. 29	72. 5. 38	70. 34. 23
25		62. 51. 46	61. 17. 54	59. 43. 36	58. 8. 51
26		50. 8. 25	48. 30. 56	46. 53. 0	45. 14. 36

Fig 7: A page of lunar distances from the first Nautical Almanac. *The small crescent stands for the moon and the dotted circle for the sun.*

With the help of these two publications the offshore navigator could now measure his longitude quickly and comparatively easily, provided he had a good Hadley quadrant (costing about £5) or, better still, a sextant (costing more like £15), and a decent watch. One or (preferably) several chronometers would make life easier still, but they would cost at least 40 guineas (£42) each—this at a time when a Royal Navy lieutenant's *maximum* annual salary was only £84.[18] Ships' officers in the East India Company had already begun to take up the use of lunars following Maskelyne's example on his excursion to St. Helena, and they warmly embraced the new publications. The Royal Navy, for its part, soon began to supply the *Almanac* and *Requisite Tables* to their ships, and in 1769 masters of all naval vessels visiting Portsmouth were ordered to obtain instruction from the Royal Naval Academy in the use of the Hadley quadrant and *Almanac*. This instruction was not well received and very few masters seem to have acted on it, but it is clear that a revolution in navigation had begun.[19]

Before the *Almanac* became available, only a handful of navigators had succeeded in measuring their longitude when out of sight of land, and according to Maskelyne's biographer its first publication was "the most important date in the history of the art of navigation, certainly since the invention of the reflecting quadrant thirty-six years earlier, perhaps since the beginnings of latitude navigation back in the fifteenth century."[20]

THE LONG-STANDING RIVALRY between the advocates of the two different approaches to finding the longitude—which goes back to the clashes between Harrison and the Longitude Board—has never entirely subsided, but it reflects a failure to grasp their crucial interdependence. No prudent navigator before the mid-nineteenth century would have dreamed of relying exclusively on a chronometer (or even several of them) to find his position during a long ocean voyage. However convenient to use, chronometers were delicate, temperamental devices, and far from perfectly accurate. Even if they ran regularly, any errors would steadily accumulate. On the other hand, if they did not run regularly they were of little use. In either case, the navigator crucially needed some independent means of determining the time to establish their rates.

As late as 1839, Norie's *Complete Epitome of Practical Navigation*—a standard manual—was offering the following advice:

> *several ingenious artists . . . have brought chronometers to an astonishing degree of perfection, whereby they have become a valuable acquisition to the navigator, in determining the difference in longitude made in short periods: however, considering the delicacy of their construction, and the various accidents to which they are liable, an implicit confidence ought not to be placed on them alone, particularly in long voyages; but recourse should be had to astronomical observations, whenever opportunities present themselves.*[21]

Another leading navigational authority, Henry Raper, who fully recognized the value of chronometers and strongly recommended their use,[22] also commented on their unreliability. In *The Practice of Navigation and Nautical Astronomy*, first published in 1840, he observed:

> *Chronometers are generally found to perform best at the beginning of a voyage; many subsequently become useless from irregularity, and some fail altogether. They are liable, also, to change their rates suddenly, and then to resume the former rates in a few days.*[23]

Lunars certainly had their shortcomings. They were tricky to observe, laborious to calculate, and often unobtainable. However, they had the crucial merit of providing a reliable and independent means of *finding* the time wherever the navigator happened to be. A chronometer, on the other hand, was always available (so long as it worked), and it enabled the navigator to *keep track* of the time during the intervals between lunar observations. In practice, the two methods were entirely complementary, and they were to be used in tandem for at least seventy years. It was only with the invention of the electric telegraph and the laying of submarine telegraph cables that lunars ceased to be essential.

Chapter 8

Captain Cook Charts the Pacific

Day 9: Wind W by S force 4–5. Not a very good day. Various mishaps: failure of engine circulating water when trying to charge batteries; the binnacle light wouldn't go on; and the washing up liquid drained away—so we now have to wash up in plain seawater. Pretty tired from night watches, which Colin and I share. Lucky Alexa is excused.

We put our clocks forward an hour at 1700. Time passes so very slowly. Anything at all complicated seems to be much harder when you're tired, especially when the boat is bouncing around. It took me more than half an hour to do the calculations for the noon fix, which finally came out as 43°16' N, 41°51' W, making a day's run of 105 miles. A really good supper—sardine pilaf made by Alexa. Still under No. 2 stays'l. This evening Colin showed me how to do a Polaris sight.

The heroic age of scientific hydrography began in the late 1760s, when the major naval powers—Britain, France, and Spain—began to send out expeditions to explore and chart the world employing the latest navigational technology.

The work of the new breed of naval surveyors—often conducted in conditions of great hardship and danger, far beyond the reach of any help—played a crucial part in the emergence of the world we know. The development of safer, faster, and more reliable transoceanic transport routes depended heavily on the accuracy of the new charts they made, and was also sometimes prompted by the discoveries recorded on them. Australia, for example, might well not have been colonized by the British had it not been for the reports brought back from Cook's first voyage. The projection of Western naval, military, and commercial power around the globe, and the huge colonial expansion that helped to fuel the Industrial Revolution, were thus heavily indebted to the work of the naval surveyors. More honorably, perhaps, the expeditions they led also provided unprecedented opportunities for scientists of all kinds, and even artists, to expand the horizons of their disciplines. The new world order—for better or worse—depended on good charts, and these in turn crucially depended on the sextant.

The French hydrographic office—the Dépôt des Cartes et Plans de la Marine—was established as early as 1720. The British were very slow to follow their example, and the Royal Navy was consequently forced to rely on French charts of the Mediterranean and the Bay of Biscay—from which vital information was sometimes deliberately withheld—during the Napoleonic Wars.[1] The Admiralty Hydrographic Office was not set up until 1795 and the first charts published under its aegis appeared in 1801. Until that time all British charts were published privately. By 1855, however, the Hydrographic Office had made up for lost time, and more than two thousand Admiralty charts covering much of the world were already available.[2] Many of the early naval surveyors have been forgotten, or are at best obscurely commemorated by the places that still bear their names. The accounts of their exploits make fascinating—and sometimes terrifying—reading, but they were (for the most part) tight-lipped professional men who rarely recorded their private feelings. It is seldom possible to assess their state of mind, but there can be no doubt that they were dedicated, independent, highly skilled, immensely hardworking, and courageous. What they were doing had never been done before. They understood that lives and fortunes would depend on the quality of their work, and that doing their duty would often expose them to great danger, but they knew they belonged to a small elite: the finest navigators of the age.

In the 1760s the map of the world was marked by many blank spaces, and even where the shape of the land was shown it was often misleading. There were, however, two subjects of especially lively geographical debate. In the first place, many were still convinced of the existence of a large and fruitful continent somewhere in the unexplored southern reaches of the Atlantic, Indian, or Pacific oceans—perhaps even spanning all three. This fabled southern continent had been the subject of speculation from the days of Magellan in the early sixteenth century, though no one had yet found any reliable evidence of its existence. Second, it was still hoped—despite many unsuccessful attempts to discover it from the eastern side—that a navigable passage might be found linking the North Atlantic and Pacific oceans either in the Arctic or possibly farther south, in more temperate latitudes.* For example, there were rumors that a channel linked Hudson's Bay in northern Canada directly with the Pacific. Beyond doubt was the fact that the vast Pacific Ocean had only so far been explored in a haphazard, piecemeal fashion and that the available Pacific charts were full of gaps and baffling inconsistencies. It was therefore not unreasonable to suppose that the "southern continent," if it existed, might lie hidden in its unexplored reaches, and that the Pacific coast of North America might reveal a new and commercially valuable trade route between Europe and the Far East.

It was against this background that, in 1768, the British Admiralty decided to send an expedition to the South Seas, under

---

* Futile attempts to find a "Northwest Passage" in the Canadian Arctic continued to waste lives until the end of the nineteenth century, by which time it was clear that the route was of no commercial utility. With the gradual retreat of the ice, that judgment may soon change.

the command of an obscure warrant officer in the Royal Navy who was soon to become a celebrity throughout Europe. Envisaged initially as a contribution to the international scientific effort to observe the second Transit of Venus of the eighteenth century (the first having taken place in 1761), the expedition had the additional task of exploring the southern Pacific in search of the fabled continent.

James Cook was not the obvious choice to command the *Endeavour*, the vessel selected to convey the British scientists to the recently discovered island of Tahiti, where their astronomical observations were to be made. There were other members of the Royal Navy who were at least as well qualified, and there was one civilian who was sure that the job was made for him—Alexander Dalrymple.[3] Though not a naval officer, Dalrymple was a persuasive character, and he had solid experience of maritime exploration and chart-making in the Far East on behalf of the East India Company. He also had the support of the Royal Society and was a devoted—almost obsessive—advocate of the southern-continent hypothesis. However, when the Admiralty flatly refused to allow him to command a naval vessel, he declined to sail merely in the capacity of an observer.[4]*

Dalrymple's rejection was Cook's opportunity. From a poor farming family in North Yorkshire, Cook had started his seafaring life at an early age on board colliers in the North Sea, only joining the Royal Navy—as an ordinary seaman—in 1755, at the unusually late age of twenty-six. His abilities were so obvious that he rose quickly, soon reaching the most senior noncommissioned rank of master, in which capacity he had responsibility—under the captain—for the navigation and general management of the ship. Cook's survey work in North America had demon-

* Dalrymple was in 1795 to have the distinction of becoming the first Hydrographer of the Navy.

strated his exceptional skills, as well as his courage and initiative. In addition to his vital services on the St. Lawrence River before the fall of Quebec, he had charted much of the coast of Newfoundland and had even seized the opportunity of an eclipse of the sun to determine the longitude of an island off its south coast. This last exploit was the subject of a paper submitted on Cook's behalf to the Royal Society in London.[5] By the time he was appointed to the *Endeavour* he had won the support of a number of powerful figures in the Royal Navy.

His new command was a tough little ship of exactly the type he had sailed in the merchant service—a flat-bottomed, bluff-bowed North Sea collier, only a little over 100 feet in length while her maximum beam (or width) was less than 30 feet: the soon-to-be-famous *Endeavour.* In addition to her crew, room had to be found for the rich young gentleman-naturalist Joseph Banks and his "suite," which included the eminent botanist Daniel Solander, two artists, a secretary, and four servants.[6] The *Endeavour*'s normal complement when serving as a collier would have been fewer than twenty men,[7] but now she would be packed with more than ninety, together with all the stores, trade goods, and equipment needed on a long voyage, as well as pigs, poultry, and a goat—not to mention Banks's two greyhounds. She carried the very latest astronomical instruments, which included a "Brass Hadley's sextant, bespoke by Mr. Maskelyne of Mr. Ramsden"—one of the finest instrument-makers of the day.[8] To describe the *Endeavour* as overcrowded when Cook at last set sail in August 1768 would be a gross understatement, and she was not very fast—slower even than the much smaller *Saecwen*—seldom managing more than 120 miles in twenty-four hours, often much less. Almost at the last minute, Cook was promoted to the rank of lieutenant—with the strictly honorific rank of captain. He was a remarkable man, and he needed to be.

In the course of his three great voyages of discovery, Cook, with sextant in hand, added more to European knowledge of the Pacific Ocean than any other single person. It was he who first recognized the kinship of the peoples inhabiting the so-called Polynesian Triangle—the vast area of sea that embraces at its extremities Hawaii, New Zealand, and Easter Island.[9] On his first voyage (1768–71), he put many islands on the map for the first time, circumnavigated New Zealand (of which, until then, only parts of the west coast were known), and went on to explore most of the east coast of what we now know as Australia. On his second voyage (1772–75), Cook sailed farther south than anyone before—on one occasion coming almost within sight of Antarctica itself—in his determined effort to prove that no habitable landmass lay in that region.

The conditions Cook and his crew endured on this second voyage were rugged, to say the least, and the dangers posed by icebergs and heavy seas were severe. Not long after leaving Cape Town in late November 1772, they ran into heavy weather, and for many days Cook's ship, the *Resolution*, another collier very like the *Endeavour*, and her consort, the *Adventure*, rode out storms in which they could carry almost no canvas. As they headed farther and farther south, freezing temperatures left the sails and rigging "all hung with icicles," and they had to pick their way cautiously through icebergs and drifting pack ice. Visibility was often reduced by fog and snow, and Cook had to issue extra clothing and warm caps to the long-suffering crew to protect them against the extreme cold. The weather conditions gave them few opportunities to make lunar observations. South of the Antarctic Circle in the longitude of 39°35' East, they encountered solid pack ice, and even Cook's determination could carry them no farther.

Having searched the southern Indian Ocean, Cook turned his attention to clarifying the confused geography of the South Pacific. After visiting New Zealand and Tahiti once again, as well as Tonga, he headed south, crossing the Antarctic Circle again on three separate occasions. This was dangerous work at the best of times, but in December 1773 the *Resolution*—now separated from the *Adventure*—had a particularly narrow escape. The ship strayed perilously close to an iceberg, and by the time Cook was called on deck the situation seemed almost hopeless. The massive island of ice was so near that the men were readying themselves to fend the ship off as best they could, but she managed—just—to get clear. One of the midshipmen later described it as "the most *Miraculous* escape from being every soul lost, that ever men had."[10] Cook dryly commented: "a miss is as good as a mile, but our situation requires more misses than we can expect. . . ."[11] On Christmas Day, while between 135 and 134 degrees West and just north of the Antarctic Circle, the crew were allowed to celebrate, but not everyone was pleased. Johann Forster—the gifted but awkward German-born naturalist who was something of a figure of fun on board the *Resolution*—gave a lurid account of the occasion:

> *The Islands of Ice surrounding the Ship look like the wrecks of a destroyed world, every one of them threatens us with impending ruin, if you add our solitary Situation & being surrounded by a parcel of drunken Sailors hollowing & hurraing about us, & peeling our Ears continually with Oaths & Execrations, curses & Dam's it has no distant relation to the Image of hell, drawn by the poets: & were it not for the pinching cold, we would really think it were still more similar.*[12]

At the end of January 1774, once again south of the Antarctic Circle, though now at 106 degrees West, Cook encountered an immense ice field that extended as far as the eye could see to east and west. The comments in his journal at this point are revealing of this extraordinary man's pride and determination, as well as his deep sense of duty:

> *I will not say it was impossible anywhere to get in among this Ice, but I will assert that the bare attempting of it would be a very dangerous enterprise and what I believe no man in my situation would have thought of. I whose ambition leads me not only further than any other man has been before me, but as far as I think it possible for man to go, was not sorry at meeting with this interruption, as it in some measure relieved us from the dangers and hardships, inseparable with the Navigation of the Southern Polar regions.*[13]

Cook could now have headed for home, but instead he crisscrossed the South Pacific, systematically verifying the existence and accurately determining the positions of many important island groups reported—often unreliably—by earlier explorers, while also making new discoveries. On his way home he briefly reconnoitered the wild southern coast of Tierra del Fuego and probed the South Atlantic, where he charted the north coast of South Georgia and discovered the even less hospitable South Sandwich Islands, lying still farther to the south. To the dismay of Dalrymple and the other promoters of the southern-continent hypothesis, his conclusion on this subject was devastatingly clear, but he also rightly took credit for his extensive survey work in the tropics:

> *I had now made the circuit of the Southern Ocean in a high Latitude and had traversed it in such a manner as to leave not the least room for the Possibility of there being a continent, unless near the Pole and out of the reach of navigation; by twice visiting the Pacific Tropical Sea, I had not*

> *only settled the situation of some old discoveries but made there many new ones and left, I conceive, very little more to be done even in that part. Thus I flater* [sic] *my self that the intention of the Voyage has in every respect been fully Answered, the Southern Hemisphere sufficiently explored and a final end put to the searching after a Southern Continent, which has at times ingrossed the attention of some of the Maritime Powers for near two Centuries past and the Geographers of all ages.*[14]

Forster again gives a vivid account of what life was like aboard the *Resolution* in the stormy waters of the deep south:

> *our Ship is tossed backwards & forwards, up & down the mountainous waves: each summit, from which you may overlook the vast extent of the Ocean, follows again a deep abyss, where we get hardly light in our Cabins. . . . At 9 o'clock, there came a huge mountainous Sea & took the ship in her middle, & overwhelmed all her parts with a Deluge. The table in the Steerage, at which we were sitting, was covered with water, & it put our candle out: the great Cabin* [Cook's] *was quite washed over & over by the Sea coming through the Sides of the Ship. Into my Cabin came the Sea through the Skuttel & wetted all my bed. I had new sheets laid & the bed rubbed up & dried as well as could be done, & in this damp bed I turned in . . . but the continual rolling of the Ship hindered me from Sleeping. . . . The Ocean & the winds raged all night.*[15]

Cook reached home in July 1775 and was greeted as a hero—rather like Neil Armstrong returning from the moon, though he had been away for much longer. He was promoted to the rank of post-captain and, early the following year, elected a Fellow of the

Royal Society. He could now have enjoyed a comfortable retirement with his wife and children, but the challenge of further exploration could not be resisted. In July 1776 Cook set sail on his final voyage with two ships—the *Resolution* (again) and the *Discovery*—this time heading into the far north of the Pacific in search of the fabled northern route to the Atlantic. On the way he became the first European discoverer of the Hawaiian Islands. Though very doubtful of the passage's existence, he dutifully reconnoitered the coasts of what are now British Columbia and Alaska, and then, having passed through the Aleutian Island chain, pressed on into the Arctic Ocean, beyond the Bering Strait, where solid pack ice at last forced him to reverse his course. On his return journey, he coasted among the Hawaiian Islands, finally anchoring at Kealakekua Bay on the west coast of the great island of Hawaii itself on January 17, 1779.

Cook was received there almost—or perhaps actually—as a god, and the ships obtained plentiful provisions, carried out necessary repairs, and made the usual astronomical observations. After a stay of almost three weeks Cook departed, but damage caused by a violent gale unluckily obliged him to return to the same anchorage to carry out further repairs. His reception on this second visit was much cooler, and his customary restraint and tact in his dealings with native peoples now deserted him. Perhaps the strain of three extraordinarily demanding voyages had at last proved too much, even for him.

On February 14 he attempted to recover a stolen ship's boat by taking a chieftain hostage—a strong-armed stratagem he had employed successfully elsewhere. On this occasion, however, he gravely misjudged the mood of the large throng that greeted him as he went ashore, and exposed himself and his small guard of marines to unnecessary danger. A confused fracas broke out during which he and several of his men were killed, as were many of the natives. Cook's companions were appalled and outraged, but his second-in-command, Lieutenant Charles Clerke, though himself fatally ill, was able to prevent reprisals that might well have led to a bloodbath. The few remains of Cook that could be recovered—most of his body had by then been divided up among the island's chiefs—were solemnly committed to the waters of the bay. When news of his death reached home he was lamented not just in Britain but throughout Europe.

READING ABOUT COOK'S exploits today, it is easy to underestimate the scale of the challenges he faced. So many of the places he visited are now holiday destinations that we may even envy him the privilege of seeing them in their pristine state—before they were ravaged by imported pests and diseases and turned upside down by missionary Christianity, colonial exploitation, and tourism. It requires an effort of imagination to grasp how demanding and relentless were the difficulties he had to overcome.

The single biggest problem, apart from the safe management of his ship, was ensuring that her crew and passengers had adequate supplies of fresh food and water—an anxious and unremitting task when sailing in almost completely uncharted waters. Less obviously, wood for the cooking fires was also vital and not always easy to obtain. Always present in Cook's mind was the tension between, on the one hand, the demands of prudent seamanship and the proper care of his men and, on the other, the need to gather as much navigational and scientific information as possible. But he was also well aware of the grave risks to which the first, mutually uncomprehending contacts between his crews and native peoples would give rise, and—to his credit—took strong measures both to prevent acts of violence and to avoid the spread of sexually transmitted diseases.

Sadly these all too often proved ineffective. "Tell me"—he asked rhetorically—"what the Natives of the whole extent of America have gained by the commerce they have had with Europeans?"[16] He knew the answer all too well.

Cook's journals, written on the spot in his own blunt style and full of misspellings, shed a vivid light on the enormous strains and hardships—both physical and psychological—involved in exploring unknown coasts far beyond the possibility of rescue in small, slow, crowded sailing ships with no effective auxiliary source of propulsion. On many occasions he and his crews came close to disaster, but seldom were they in greater danger than when they were carrying out a "running survey" of the east coast of Australia on his first voyage. Cook chose not to explain his survey methods in detail, but this process entailed following the coast as closely as possible during the hours of daylight, taking compass bearings of all prominent landmarks, sounding continually with the lead, and all the while keeping close track of the ship's course and speed. "Ship stations" (fixed points) would have been established from time to time by taking horizontal sextant angles between prominent landmarks. Cook and his officers—as well as the artists—also made sketches of significant coastal features whenever they could.

Having found and named Botany Bay, where Banks and the other "scientific gentlemen" were delighted by the discovery of hundreds of new plant and animal species, Cook sailed north into the dangerous waters that lie between the Great Barrier Reef and the mainland. No European ship had been there before, and the *Endeavour* was alone: she had no consort on which to rely in case of difficulty. This was a mistake that Cook would not repeat on his two subsequent voyages.

On the night of June 10–11, 1770, when the ship was just north of Cape Tribulation (aptly named by Cook himself "because here begun all our troubles") in what is now Queensland, Cook cautiously shortened sail and stood out from the mainland. His aim was to avoid dangers visible ahead and to confirm whether there were any islands in the offing, but he was as yet unaware of the existence of the Great Barrier Reef, which now hemmed him in ever more tightly. There was a good sailing breeze and it was a "clear moonlight night"; soundings were taken continuously with the lead-line. The water began to shoal, and the crew were anxiously standing by to drop anchor when it suddenly became deeper again. Cook decided he could risk carrying on, but just before 11 P.M., moments after a sounding had shown they were in 17 fathoms (102 feet), "and before the Man at the lead could heave another cast the Ship Struck and stuck fast."[17]

The *Endeavour* had grounded on an isolated coral reef at the top of the tide—the worst possible moment, as it meant the sea level would soon be falling—and she was badly holed. It was an extremely perilous position, so desperate in fact that Cook ordered the guns, ballast, stores, and even freshwater to be thrown overboard to lighten the ship. The crew carried anchors out in the boats and tried to haul her off, but she could not be moved. Luckily the wind was light, so the *Endeavour* was not pounded to destruction, as she well might have been, and after twenty-three hours of exhausting work they at last succeeded in getting her afloat again at the top of the next high tide but one. However, the damage to the hull was so extensive that it was far from clear they would be able to reach the shore (about twenty miles away). By "fothering" the ship—passing a sail coated with a mixture of wool, rope yarn, and dung underneath the damaged part—they managed to get the massive leak under control, and eventually on June 16 succeeded in beaching the *Endeavour* in a river mouth on the mainland. Cook had chosen his ship well: as a collier, she was designed to take the ground for the discharge of her cargo without suffering damage.* A sizable chunk of coral was found jammed in the hull; if it had been dislodged while they were still at sea the ship would almost certainly have been lost. Banks was deeply impressed by the "cool and steady conduct" of the officers, "who during the whole time never gave an order which did not show them to be perfectly composed and unmov'd by the circumstances however dreadful."[18]

While the ship was being repaired, Cook and Green (the professional astronomer who accompanied him) managed on June 29 to obtain a very accurate longitude by observing the first moon of Jupiter: 214°42'30" West (or 145°17'30" East).[19] Having set sail again on August 4 it proved extremely difficult to find a safe passage through the "labyrinth" of reefs that extended far beyond the river mouth. When at last they reached the open sea, Cook—though sorry to lose the chance to survey the mainland shore—was relieved at being outside the shoals in which he had been "intangled" since the end of May:

> *in which time we have saild 360 Leagues without ever having a Man out of the cheans† heaving the Lead when the ship was under way, a circumstance I dare say never happen'd to any ship before and yet here it was absolutely necessary.*[20]

But Cook and his crew were far from safe. On August 16, while sailing northward just outside the Great Barrier Reef, the *Endeavour* was becalmed during the night in a heavy swell that began to drive her slowly but irresistibly toward it:

---

* Larger vessels, like the frigate employed by Bougainville, though faster and bigger, would probably have broken their backs if beached.

† That is, the "chains"—see Glossary.

A little after 4 oClock the roaring of the Surf was plainly heard, and at day break the vast foaming breakers were too plainly to be seen not a Mile from us, towards which we found the Ship was carried by the waves surprisingly fast. We had at this time not an air of wind, and the depth of water was unfathomable so that there was not a possibility of Anchoring. . . . The same Sea that washed the side of the Ship rose in a breaker prodigiously high the very next time it did rise, so that between us and destruction was only a dismal Valley the breadth of one wave, and even now no ground could be felt with 120 fathoms.[21]

They were at least thirty miles from the nearest land, and there were not enough boats to carry the whole crew, yet in this "Truly Terrible Situation not one man ceased to do his utmost, and that with as much Calmness as if no danger had been near."

All the dangers we had escaped were little in comparison of being thrown upon this Reef where the Ship must be dashed to peices in a Moment. A Reef such as is here spoke of is scarcely known in Europe, it is a wall of Coral Rock rising all most perpendicular out of the unfathomable Ocean . . . the large waves of the vast Ocean meeting with so sudden a resistance make a most terrible surf breaking mountains high especially as in our case, when the general trade wind blowes directly upon it.*[22]

According to Banks, the situation was so desperate that "a speedy death was all we had to hope for."[23] At this critical moment, how-

* It is worth pointing out that the trade winds are not gentle zephyrs: they often blow hard. The great advantage they offered in the days of sail was their regularity and predictability.

ever, a breath of wind, so small that it would have passed unnoticed in normal circumstances, enabled them to edge away from the reef—with additional help from the oarsmen sweating in the ship's boats. Soon, however, the "Friendly breeze" failed them and they were once again in danger.

> *A small opening was now seen in the Reef about a quarter of a Mile from us which I sent one of the Mates to examine, its breadth was not more than the length of the Ship but within was smooth water, into this place it was resolv'd to push her if possible, haveing no other probable Views to save her, for we were still in the very jaws of distruction, and it was a doubt whether or no we could reach this opening, however we soon got off it when to our surprise we found the Tide of Ebb gushing out like a Mill Stream so that it was impossible to get in; we however took all the advantage possible of it, and it carried us out about a ¼ of a Mile from the breakers.*[24]

By noon they were a mile or two off the reef but were certainly not out of danger. Cook therefore sent Lieutenant Hicks in a small boat to explore another opening in the reef that they could see a mile or so to the west.

> *At 2 oClock Mr Hicks returnd with a favourable account of the opening, it was immediately resolved to try to secure the Ship in it, narrow and dangerous as it was it seem'd to be the only means we had of saving her as well as our selves. A light breeze soon after sprung up at ENE which with the help of our boats and a flood tide we soon enter'd the opening and was hurried through in a short time by a rappid tide like a Mill race which kept us from driving*

> *against either side, though the C[h]annell was not more than a quarter of a Mile broad, we had however two boats a head to direct us through. . . .*[25]

The *Endeavour* was now out of danger, and Cook—in an uncharacteristic departure from his normal tight-lipped stoicism—took the opportunity to comment on the extraordinary stresses that accompanied the work in which he was engaged. Coming from perhaps the most revered navigator in the history of exploration, the anxieties—and the note even of bitterness—that emerge from this powerful testimony are particularly telling:

> *It is but a few days ago that I rejoiced at having got without the Reef; but that joy was nothing when Compared to what I now felt at being safe at an Anchor within it. Such are the Visissitudes attending this kind of Service, and must always attend an unknown Navigation where one steers wholy in the dark without any manner of Guide whatever. Was it not from the pleasure which Naturly results to a man from his being the first discoverer, even was it nothing more than Land or Shoals, this kind of Service would be insupportable, especially in far distant parts like this, Short of Provisions and almost every other necessary. People will hardly admit of an excuse for a Man leaving a Coast unexplored he has once discovered. If dangers are his excuse, he is then charged with Timerousness and want of Perseverance, and at once pronounced to be the most unfit man in the world to be employ'd as a discoverer; if, on the other hand, he boldly encounters all the dangers and Obstacles he meets with, and is unfortunate enough not to succeed, he is then Charged with Temerity, and, perhaps, want of Conduct.*[26]

Cook plainly did not expect to incur these criticisms, but he acknowledged that he had perhaps been less than prudent in penetrating so deeply among the islands and shoals on this stretch of coast "with a single Ship, and every other thing considered." On the other hand, if he had not taken these risks he would have fallen short of his own high standards:

> *. . . I should not have been able to give any better account of the one half of it than if I had never seen it; at best, I should not have been able to say wether it was Mainland or Islands; and as to its produce, that we should have been totally ignorant of as being inseparable with the other; and in this case it would have been far more satisfaction to me never to have discover'd it, but it is time I should have done with this Subject wch at best is but disagreeable & which I was lead into on reflecting on our late Danger.*[27]

Cook does not mention—perhaps he did not even notice—that in the middle of this crisis Charles Green, the professional astronomer who accompanied him, was hard at work making lunar observations to help fix their position. Green's personal log records in neat copperplate script:

> *These obs[ervations] were very good, the Limbs of sun and moon very distinct, and a good Horizon. We were about 100 Yards from the Reef, where we expected the Ship to strike every minute, it being Calm & no soundings, and the swell heaving us right on.*[28]

Green was not a sailor by training (though as Maskelyne's assistant he had sailed with him to Barbados for the trials of H4 and held the naval rank of purser), but he was a cool, determined hand, and Cook described him as an "indefatigable" observer.

In fact he seems to have taught Cook how to make lunar-distance observations while on the outward passage from England.[29] That anyone would attend to lunars at such a time is a clear demonstration of their vital importance, and of Green's devotion to duty. His log reveals that he was responsible for the majority of the lunar observations made on the voyage, though Cook's name quite often appears, too. It was a heavy responsibility, and the "young gentlemen" frequently maddened him by failing to provide the assistance he expected. Like a disappointed teacher, Green plainly found it difficult to understand why they did not share his enthusiasm for the higher flights of celestial navigation (Green's own emphasis):

> *22 September 1768— . . . might have made more and better [observations], if Proper Assistance could have been had from the young Gentlemen on board, with pleasure to themselves . . .*[30]

> *24 February 1769—[The sea] broke over the Quarter Deck several times while we were observing. . . . The Moon's altitude I took myself . . . because I could get no one to assist in taking it for me.*[31]

Green was never to return home. After picking his way through the hazardous and hitherto uncharted Torres Straits, Cook was obliged to call at the appallingly unhealthy Dutch settlement of Batavia (modern Jakarta) to make essential repairs to the ship. Malarial fever and dysentery ran through the *Endeavour*'s crew, and Green was among the many victims, dying at sea shortly after they set sail for home in January 1771. By the time they reached Cape Town the death toll had reached thirty-four, and five more were to succumb before they reached home—more than a third of the ship's original complement.[32] They had survived shipwreck on the Great Barrier Reef and had reached Batavia in remarkably good health, so it was a cruel irony that this detour to an outpost of European civilization should have cost so many lives.

LUNARS WERE ALL the more vital on Cook's first voyage because he was not equipped with a timekeeper, though by that time—as both he and Green would have known—H4 had proven its worth. They did, however, have the benefit of the new *Nautical Almanac*, though its predictions expired before the long voyage was over. (With such extended voyages in mind, Maskelyne was to ensure that future editions of the *Almanac* were produced at least four or five years in advance.)

When Cook set sail again in 1772 aboard the *Resolution*, this time accompanied for safety by the *Adventure*, four timekeepers were carried, though only one of them performed reliably enough to be of outstanding navigational value. This device was a copy of Harrison's H4, specially commissioned by the Admiralty from Larcum Kendall, which was now known as "K1." Cook had high praise for K1, which he described as "our trusty friend the Watch,"[33] but he certainly did not rely on it exclusively. Accompanied on this voyage by the astronomer William Wales (who was Charles Green's brother-in-law and later taught Samuel Taylor Coleridge, the future author of *The Rime of the Ancient Mariner*), he took every opportunity to determine the longitude by lunars. Much of the value of K1 lay in accurately carrying forward the time from one set of lunar observations to the next, and in "reducing" the longitude of one place to that observed at another. This important process involved comparing the local time observed at a new location with the Greenwich time carried forward by the "watch" from the last place where a good set of lunars had been obtained. By this means the longitude of different locations could be established without the need to take further lunar observations.

When exploring the New Hebrides archipelago in the South Pacific on this second voyage, Cook listed the extensive lunar observations made by Wales (amounting to several hundred) to fix the longitudes of two ports on separate islands. These, he explains,

> *have been reduced by means of the Watch to all the islands, so that the Longitude of each is as well assertained as the two ports above mentioned, as a proof of this I shall only observe that the difference of Longitude between the two Ports pointed out by the Watch, and by the observations did not differ from each other two miles. This also shews to what degree of accuracy these observations are capable of, when multiplyed to a considerable number, made with different Instruments and with the Sun and Stars on both sides of the Moon. . . . If we consider the number of observations that may be obtained in the course of a Month (if the weather is favourable) we shall, perhaps, find this method of finding the Longitude of place as accurate as most others, at least it is the most easiest to put into practice and attended with the least expence to the observer.*[34]

After discussing the importance of investing in a sufficient number of good "quadrants"—a term that, in his usage, embraced the sextant—Cook offers the following opinion on watches:

> *The most expensive article, and what is in some measure necessary in order to come at the utmost accuracy, is a good watch; but for common use, and where the utmost accuracy is not required, one may do without.*[35]

Cook had no time for the widely held view that lunars were too demanding for the ordinary naval officer:

> *this method of finding the Longitude is not so difficult, but that any man with proper applycation and a little practice may soon learn to make these observations as well as the astronomers themselves. I have seldom found any material difference between the observations made by Mr Wales and those made by the officers at the same time.*[36]

The naturalist Johann Forster may not have been a sailor, but he was a highly intelligent and erudite man, and his comments when the *Resolution* arrived in New Zealand are of interest in this context:

> *. . . I must do justice to our Navigators, who, when we first made the Land, coincided within a few Miles with their accounts, since the last Observation of Longitude: some were not above 3 miles out. This proves more & more the Excellence of this Method of Observing the Longitude by Distances of the Moon from the Sun or some fixed Stars, which has been encouraged by the Board of Longitude & is now so well understood & practised by the Gentlemen of the British Navy.*[37]

On the return leg of this voyage, Cook measured the difference in longitude between St. Helena (accurately determined by Maskelyne in 1761) and Cape Town (established in the same year with equal care by Mason and Dixon, the British surveyors who later made history by drawing the line dividing Pennsylvania from Maryland) using both lunars and the watch. Comparing the results with the accepted figure, he found that K1 was in error by 2 miles, while the lunar observations made by Wales were out by just 5—effectively a tie. Given that the longitudinal displacement between the two sites is roughly 24 degrees, and the great circle distance is some 1,800 nautical miles, such tiny discrepancies were insignificant. Cook's enthusiasm for lunars remained undiminished: "I mention this to shew how near the longitude of places may be found by the lunar method, even at sea, with the assistance of a good watch."[38]

The most important characteristic in a chronometer is that its rate be regular, and therefore predictable: absolute accuracy is neither expected nor required. In this respect K1 was outstanding, constantly gaining between nine and thirteen seconds a day from the *Resolution*'s arrival in New Zealand in April 1773 until she reached home again in July 1775. In a letter to the Admiralty sent from the Cape of Good Hope on the return voyage, Cook reported that Kendall's watch had "exceeded the expectations of its most Zealous advocate and by being now and then corrected by Lunar observations has been our faithful guide through all the vicissitudes of climates."[39] K1 accompanied Cook on his third voyage and again performed very well, though two months after Cook's death it stopped and did not again run reliably until Kendall himself had overhauled it.

Kendall was paid the enormous sum of £450 to make K1[40]—and it took him two years to complete. He was also given a £50 reward. For comparison, the total purchase price of the *Endeavour* was only £2,800.[41] As designs and production methods improved, the price of chronometers gradually came down (by the end of the eighteenth century they cost between 60 and 100 guineas—£63 and £105—though to this should be added five or ten guineas a year for maintenance), but they remained beyond the pocket of all but the richest officers until well into the nineteenth century. It was to be many years before the use of chronometers became commonplace. Bearing in mind its customary parsimony, it is not surprising that by 1802 the British Admiralty had supplied only 7 percent of Royal Navy vessels with chronometers at official expense.[42]* These would all have been deployed aboard ships destined for service in distant waters, and notable among them were the survey vessels. Several were commanded by men whom Cook himself had trained, and they in turn trained the next generation. The small world of hydrography was a tight-knit one, and it was governed by a kind of apostolic succession.

---

* Even as late as 1821 the Admiralty owned only 130 chronometers, and it was not until 1859 that all captains' commands were required to carry three chronometers. I am grateful to Richard Dunn of the National Maritime Museum for this information.

Chapter 9

Bougainville in the South Seas

Day 10: Weather deteriorating. Up as usual at 0400, in pouring rain and heavy squalls which come creeping up under low black, fast-moving clouds. Wind W by S force 6–7 gusting to 8. At 0730 the wind eased slightly and I put up the No. 1 stays'l and, of course, it immediately blew up again so I had to change back to the No. 2. Repeated the same procedure soon afterwards. With a following wind it's very hard to choose the right amount of canvas—either too much or too little. Also the boat rolls abominably. Low dark ragged clouds above us, racing past us. No sights today.

Later that morning the wind increased to gale force, and we switched down to the much smaller storm jib. Even so, our speed through the water increased and we were surfing down steep, breaking waves that grew bigger all the time. The speed of the change was startling: in the space of an hour or two, our world had been completely transformed. I was steering by hand now, as the self-steering gear could not cope with the heavy following seas that were advancing across the surface of the ocean at a much faster pace than ours. As each wave came up behind us, the stern lifted, the bows dipped, and *Saecwen* accelerated down its face, until the crest passed beneath us, at which point we slid stern-first into the trough behind it. Sometimes the wave started to break just as it reached us and we steered to avoid the mass of hissing foam as it roared past.

Saecwen was now going very fast, perhaps too fast for safety, throwing up a huge bow wave on either side as she surfed down the waves. It was exhilarating, and I felt as if I were at the helm of a dinghy rather than a heavy yacht. Colin appeared from down below with a tense look on his face I had not seen before. He asked me to douse the jib and took the helm while I went forward to deal with it. I freed the halyard, pulled down the sail, undoing the hanks that clipped it to the forestay, and gathered it, kneeling on the foredeck, which was leaping and plunging in the heavy seas. As the bows dipped sharply downward I was almost weightless, but my knee collided painfully with the anchor windlass when the deck suddenly heaved up again. A moment later, a cascade of cold water poured up inside my oilskin trousers, soaking me to the waist. I yelled in rage and frustration as I struggled to control the sail, but luckily Colin could hear nothing above the noise of the wind and waves. I hugged the wet sail to me and staggered slowly back to the cockpit.

We were now running under bare poles, and though our speed had slackened, the weight of the wind on the mast and rigging was still driving us through the water at five or six knots. It was blowing a full gale, and the rain and spray had reduced the visibility to maybe half a mile. I went below and got out of my wet clothes. There was no chance of drying anything, so I just squeezed them out onto the cabin floor—which was already running with water—and chucked them into the oilskin locker to wait for a sunny day. My knee was bleeding and I felt very sorry for myself. Down below, the roar of the waves and the howl of the wind were slightly muffled, but the wild swooping and rolling of the boat had pried loose everything that would move. The cutlery was crashing around in its drawer, plastic mugs and plates were jiggling in their racks, cans and bottles carefully stowed in various corners with towels and socks to keep them quiet were now chinking, and several books had jumped out of the bookcase onto the floor. Alexa emerged from her bunk to help me tidy things up and put on the kettle to make hot chocolate. Colin looked tired as I handed up a mug to him. Lunch consisted of stale crackers with pâté from a tin, syrupy tinned peaches slurped out of mugs, and Mars bars. The warmth of the food and drink spread through me and I climbed gratefully into my clammy sleeping bag, to doze rather than sleep, still half aware of the motion of the boat and the swishing and gurgling of the sea running past the curved hull just a few inches away.

But sleep must have come because I dreamed. I was a small child again, flying high, as free and strong as an eagle, far above an endless sea of clouds that glowed in the golden light of the setting sun. Suddenly I remembered myself and began to fall, faster and faster, until wild waves rushed up toward me. I tried to call out but no sound came. With a crash I hit the water and sank into the dark, knowing that I was dying and would never see the light of day again. Then a big, warm hand gripped mine and pulled me up to safety. I recognized that hand at once: it was my father's.

Colin called me later in the afternoon, and I struggled reluctantly into my still-wet oilskins and squelching boots—a surprisingly difficult task when being thrown around the cabin by each lurch and roll of the boat. The scene that confronted me when I came on deck was awe-inspiring. The wind had probably not strengthened much, but the waves had reached majestic proportions. Their crests were perhaps one hundred yards apart, and the wind eased noticeably as we dipped into the shelter of the deep troughs between them. Colin was wearing his safety harness and I put mine on, too, clipping it to a strong point in the cockpit. When he went below Alexa came up to keep me company. Neither of us had ever seen such a sea before, and this was not even a big storm—just a typical North Atlantic gale. It was beautiful, exhilarating, and frightening. There was a lot of noise: the wind whistling through the rigging, the roaring hiss of the breaking waves around us, and the creaks and groans of the boat herself. We were right out in the middle of the ocean now, a long, long way from any help. I had to grip the tiller with both hands to keep *Saecwen* on course, and for the first time I felt it vibrate as the rudder cut through the water beneath us. There was no question of taking any sights that day as the sky was completely obscured by thick, low, fast-moving clouds, but at least we were more or less on the right course.

While Alexa and I were together on watch, a really big breaking wave caught us, smothering the cockpit in foaming water and casually flinging *Saecwen* down on her beam ends. Alexa was now at the helm, and hung on desperately to the tiller as the wall of cold water crashed over us, but she could exert no control whatever. The boat broached to, slewing round through ninety degrees, and for a few moments her masthead almost touched the sea before the weight of her keel slowly righted her. I found myself up to the waist in water in the cockpit, which was completely awash. Having lost her grip on the tiller, Alexa had almost gone over the side and had been badly bruised by the guardrail, but although the canvas spray-hood over the main hatch was torn we were still afloat and—thank God—the mast was still standing. While the water gradually drained from the cockpit I struggled to get *Saecwen* under control before another wave hit her. The main hatch slid back and Colin's head appeared: "Everything all right, Mr. Mate?" he asked. Alexa and I laughed nervously, but things were getting too exciting, so now we lay ahull, with no sails up, the helm lashed to one side, very slowly reaching across the track of the advancing waves rather than running fast before them. In this configuration *Saecwen* could be left to her own devices, even though the occasional comber swept with a roar over her decks, briefly darkening the cabin.* Since the visibility was so poor, we all stayed below and prepared a fortifying supper: corned-beef hash made with instant mash, onions, and eggs, followed by tinned rhubarb and then whisky. It was a wild night and we only peeped out of the hatch occasionally to make sure that all was well, but we were all in surprisingly good spirits.

THE BRITISH WERE not alone in exploring the Pacific Ocean and its shores during the latter half of the eighteenth century. Following their defeat in the Seven Years' War, the French were determined to restore their dented national pride and to play their part in extending the scope of geographic knowledge. Two great French navigators left their marks on the map of the Pacific, though their experiences, and their reactions to them, were markedly different.

Louis-Antoine de Bougainville (1729–1811)—absurdly best known today in the English-speaking world for the tropical plant that bears his name—set sail from France in November 1766 in the frigate *La Boudeuse* ("the sulky"), almost two years ahead of Cook. Unlike Cook, he was a relative newcomer to the sea. He had started his career on the staff of the French ambassador in London (where he met Anson) and later distinguished himself in the army as aide-de-camp to General Montcalm.[1] A highly educated and cultivated man, he had, while still in his twenties, published a two-volume treatise on integral calculus and was already a Fellow of the Royal Society in London. With feigned modesty, Bougainville warned readers of his *Voyage*

---

* This tactic suited *Saecwen*—with her long, heavy keel and low freeboard—well. Light-displacement modern yachts with high topsides might run the risk of being rolled over and dismasted.

autour du monde that his style was "all too plainly" marked by the wild, nomadic life he had for so long been leading. "It is neither in the forests of Canada nor on the breast of the sea that one develops the art of writing,"[2] he proclaimed.

In fact, his *Voyage* is beautifully crafted, and among the reading public it found an appreciative audience. It even prompted a reaction from the great encyclopaedist Denis Diderot, whose satirical *Supplement to the Voyage of Bougainville* (1772) was influential in spreading the Romantic notion of the "noble savage." However, Bougainville's achievements were much less impressive from a purely navigational point of view than those of Cook,[3] and his *Voyage* made only a slight impression on the world of scholarship.[4]

Bougainville was the first explorer to be accompanied by a suite of scientists—as well as an artist—an example followed by many of his successors, including Cook. His astronomer, Pierre-Antoine Véron, did not enjoy the benefit of a chronometer, but he was equipped with the latest lunar-distance tables in the *Connoissance des Temps* (for the moon and sun) and the Abbé de La Caille's catalog of fixed stars.[5] Although Véron had brought with him an instrument called a "megamètre" for making lunar observations, Bougainville judged that Hadley's quadrant was in general preferable.[6] In addition to these working "supernumeraries," Bougainville also invited a paying passenger, the Prince de Nassau-Sieghen, a young man rumored to have been involved in a number of amorous liaisons who had also run up an embarrassing number of debts. They seem to have got on well with each other. In fact, both men were enthusiastic gamblers, and they may well have been lovers of the same famous actress and opera singer, Sophie Arnould.[7] Bougainville and Cook could hardly have been more different.

Having called at Buenos Aires and Montevideo, Bougainville visited the Falkland Islands—or Les Malouines, as they were known to him—in order to hand over to the Spanish, in accordance with the terms of a recent treaty, a colony of displaced French Canadians he had established there a few years earlier. Having returned to Rio, where he was joined by the supply ship *L'Étoile* ("the star"), Bougainville entered the Straits of Magellan, the narrow passage linking the South Atlantic and Pacific oceans just north of Cape Horn. The weather was miserably cold, wild winds impeded their progress, the anchorages were often insecure, and it was rarely possible to make observations for longitude. The refined Bougainville found the native peoples depressingly primitive—and unbearably smelly.[8] The accidental death of a young native boy who ate some of the glass trade goods the French had brought with them deepened his sense of gloom.[9] In late January 1768, however, having spent fifty-two days struggling through a channel only 330 nautical miles long, he was delighted at last to see the open Pacific.[10]

In April 1768, Bougainville reached Tahiti,[11] quite unaware that Captain Samuel Wallis of the Royal Navy had called there eight months earlier in the *Swallow*.* It could hardly have differed more from Tierra del Fuego. He was entranced by the physical beauty of the island and its inhabitants, and fascinated by the sexual freedom they enjoyed. Even before they had anchored, the two ships were surrounded by canoes filled with islanders "giving a thousand signs of friendship" and "demanding nails and ear rings." Bougainville considered the women to be at least as beautiful as their European counterparts, and, as he breathlessly observed, "most of these nymphs were naked."[12] The Tahitian men, for their part, pressed the visitors to choose a woman, and "their gestures made it quite clear how we should make their acquaintance." Bougainville asks rhetorically how

---

* Carteret's purser, by the name of Harrison though no relation of the watchmaker, had fixed its longitude by lunar distances.

he could possibly be expected, in such a situation, to keep four hundred young French sailors at their work.

> *Despite all the precautions that we could take, a young girl came aboard . . . [she] negligently let slip a shift that covered her, and made the same impression on everyone as Venus revealing herself to the Phrygian shepherd:* she had the heavenly body of the goddess.*[13]

Needless to say, this vision caused quite a stir among the sailors, but it was Bougainville's cook who managed to slip away—though the adventure proved less enjoyable than he must have hoped. Having reached the shore with the girl of his choice, he was immediately surrounded by a crowd of "Indians" who undressed him and inspected him all over. Not surprisingly, he was terrified, as he had no idea what the natives who were excitedly examining him intended. Once they had satisfied their curiosity, however, they gave him back his clothes and the contents of his pockets. They then pressed him to approach the girl and "satisfy the desires that had led him to come ashore with her." Their efforts were in vain. The islanders eventually returned the poor cook to his ship, where he fully expected to be severely punished, but this prospect, he told Bougainville, was much less frightening than the experience he had just been through.[14] The cook was not the only one to be disappointed. A native chief made the prince an offer he could not politely refuse: he sent one of his wives to sleep with him. As the amused Bougainville ungallantly noted, she was "fat and ugly."[15]

A bizarre incident occurred during this visit to Tahiti. A young

---

* A reference to the Judgment of Paris. Dressed as a Phrygian shepherd, the Trojan prince Paris had the awkward task of choosing which of three goddesses—Hera, Athena, and Aphrodite—was the most beautiful, with dire consequences for Troy.

servant of the naturalist Philibert de Commerson went ashore with him and was—according to Bougainville—instantly surrounded by natives shouting that "he" was a woman. To protect her from their advances she was quickly returned to the ship. Her figure, voice, and behavior had already given rise to strong suspicions among the crew that she was not what she seemed, but it is curious that the Tahitians should instantly have seen through her disguise.

When Bougainville interviewed her she burst into tears, claiming to be an orphan and to have tricked her master into taking her with him on the voyage by pretending to be a man. Her reason for doing so was that the proposed voyage had piqued her curiosity.[16] Her name was Jeanne Baret (or Baré), but she was, it seems, attached to Commerson long before the ship had sailed from France. After her unmasking, Bougainville took steps to protect her, but dropped her and Commerson off on the Isle de France (Mauritius) on his way home, perhaps to avoid a scandal. Later the intrepid Baret returned to France, thereby gaining the distinction of being the first woman to circumnavigate the world. She married and was to receive a royal pension for having "shared the travails and perils of [Commerson] with the greatest courage" and for her "very discreet conduct."[17]

In describing Tahiti, the cultured Bougainville could not resist reaching for comparisons from classical literature, fine art, and even the Bible. When a fine-looking islander sang to them to the accompaniment of "a tender air" played on the nose flute, he judged that the song was "without doubt anacreontic," while the whole scene was "worthy of Boucher's brush."[18]* Walking in the gloriously verdant interior of the island, he felt as if he

* Anacreon (582–485 BCE) was an ancient Greek lyric poet. François Boucher (1703–70) was a famous painter of pastoral scenes often peopled with naked nymphs and goddesses.

had been transported to the Garden of Eden. On Tahiti, Venus was "the goddess of hospitality," though here her cult admitted of no mysteries.[19] Appropriately enough, he named the island "Nouvelle-Cythère" (New Cythera), after the legendary birthplace of the goddess.[20]

Bougainville did not, however, conceal the unpoetic fact that relations between the European visitors and the natives sometimes broke down. Thefts gave rise to many misunderstandings[21] and, much to his dismay, some natives were killed.[22] He was also aware that the inhabitants of the different islands engaged in brutal warfare and even human sacrifice.[23] Nevertheless, Bougainville was very favorably impressed by the manner in which Tahitian society was organized. It was not just that their free and easy sexual habits appealed to him; he noted approvingly that men and women shared the burdens of childcare, while the basic necessities of life were held in common.[24] To make the idyll complete, the climate was delightful and there were, as far as he could tell, no troublesome insects or venomous animals.

Bougainville was also surprised by the skill with which the islanders built their canoes and by their extraordinary navigational skills:

> *the learned people of this nation, without being astronomers . . . understand the daily movements* [*of the stars*] *and make use of them to find their way in the open sea from island to island. In making these voyages, sometimes of more than three hundred leagues, they completely lose sight of land. Their compass is the course of the sun by day, and the position of the stars during the night, which is almost always fine between the tropics.*[25]

The sensation that Bougainville's enthusiastic published account of life in Tahiti caused was reinforced by the appearance in Parisian society of Aotourou—the young, high-caste Tahitian who sailed back to France with him in 1769.

Bougainville's stay in Tahiti was no holiday, however. The ships' anchor cables, being made of hemp rather than chain, were frequently severed by the sharp coral of the anchorage, and this very nearly led to the destruction both of *La Boudeuse* and *L'Étoile*.[26] The prospect of being marooned—even in such a paradise as Tahiti—was very alarming, while the loss of six of his anchors meant that Bougainville was unable to explore the island and its neighbors as thoroughly as he wished. Véron also had problems: after making a large number of lunar observations[27] over four days and nights in order to fix the longitude of Tahiti, his precious notebook was stolen. He was, however, able to determine the position of the French camp to his satisfaction by means of eleven observations taken on the eve of their departure.[28]

Bougainville's voyage after Tahiti took him almost as far as the east coast of Australia. He encountered on June 4, 1768, a low, sandy island covered with birds, which he named La Bâture de Diane (Diana's Bank).[29] The following day some members of the crew thought they could see land to the west, but Bougainville believed they were mistaken. During the night, however, the appearance of driftwood and fruit coupled with the fact that the swell had subsided led him to suspect the presence of land somewhere to the southeast.[30] Continuing their westerly course they encountered another reef on June 6, and some thought they saw land to the southwest of the breakers. So Bougainville altered course to the north until 4 P.M., when he turned west again. At 5:30, lookouts at the masthead could see more white water to the northwest, which extended as far as the eye could see. The furious breaking of the sea on these reefs struck Bougainville like "the voice of God"[31] and rendered him "docile." Realizing that he was somewhere off the still unknown east coast of "New

Holland," he decided—cautiously but perhaps wisely in view of Cook's subsequent experience—that it was too risky to close the land.

Since Bougainville and his crew were critically short of bread and vegetables, and their salt meat was so heavily infested with maggots that they preferred to eat the ship's rats,[32] they now needed to reach one of the Dutch settlements in the East Indies—and as quickly as possible. Unsure whether a navigable passage existed in that direction, Bougainville chose not to head west along the south coast of New Guinea and across the Gulf of Carpentaria. He thus missed the chance of anticipating Cook's rediscovery of the strait that Torres had passed through in 1606. Becalmed off the south coast of New Guinea—on whose sweet-smelling and fertile shores he and his hungry crew were unable to land—they narrowly escaped shipwreck. "Terrible" days followed as they struggled to the southeast in the face of strong headwinds, rain, and even thick fog. Once again they were almost wrecked and then found themselves in a heavy sea surrounded by so many reefs that they could only close their eyes and hope. So shallow was the water that bottom-dwelling fish were washed aboard, along with sand and seaweed; there was simply no point in taking soundings. On June 16 the weather improved, but they still faced a long, hard struggle to reach the eastern end of the New Guinea archipelago. Their food supplies had almost run out and "the most cruel of our enemies, hunger, was now aboard."[33] Rations were reduced; a favorite goat and a pet dog were sacrificed. Bougainville even had to issue orders against eating the leather that protected the spars and rigging.

At last on June 26 they rounded the end of Rossel Island that marks the eastern extremity of the reef-hedged New Guinea archipelago. They called the high headland Cape Deliverance: "a piece of land that we have well earned the right to name,"[34] wrote Bougainville. He then warily sailed north, passing among the long-lost Solomon Islands, one of which is named in his honor. Like Carteret, who had recently passed this way in the *Swallow*, Bougainville had no idea that these islands were the same as those that Mendaña had discovered two hundred years earlier: understandably, as contemporary charts based on the wildly underestimated longitudes recorded by the Spanish explorer placed the Solomons much farther to the east. On July 13, 1768, observations of a solar eclipse made by Bougainville's astronomers in New Britain at last made it possible to determine accurately the width of the Pacific Ocean—"hitherto so uncertain."[35]

Bougainville's course then took him west along the north shore of New Guinea, across the Celebes Sea and then south through the Macassar Strait—regions already well-known to the Dutch. By now, scurvy was affecting almost all of his men, and he was forced to make two stops in Dutch-controlled harbors to restore their health, before heading home via Mauritius and the Cape of Good Hope. As he himself conceded, his circumnavigation yielded fewer discoveries than Cook's first voyage[36] (which partially coincided with it), but it did much to excite interest in the Pacific islands and their peoples. It also made Bougainville a celebrity. Sixteen years later an even more ambitious French expedition set sail for the Pacific, this time under the command of Jean-François de Galaup, comte de La Pérouse. The contrast between Bougainville's experiences and those of La Pérouse could hardly be more pronounced.

Chapter 10

La Pérouse Vanishes

Day 11: Hardly slept at all. Colin made inedible porridge for breakfast with far too much salt. Worked out that we were approaching the halfway mark. We're almost a thousand miles from land in rough seas under a cloudy rain-filled sky. God I wish we were home.

Westerly gale-force winds continued throughout the night as we lay ahull. At about 0800 we were all sheltering down below when we heard the loud, deep blast of a ship's horn. Rushing up on deck I saw a ship looming out of the murk only half a mile astern, and heading right for us. She was pitching heavily in the big seas. A 25,000-ton bulk carrier, according to Colin; we tried to raise her on the radiotelephone, but had no luck. Going very slowly, she circled round us, some of her crew lining the rail and waving.

We waved back, though we must have looked a sad sight, rolling around in a heavy sea with no sails up. Colin tried to show that we were okay by pretending to take swigs from a whisky bottle, though maybe this had the opposite effect. She was the *Janetta*, and her port of registry was in Norway. With her engines going astern, she ranged up close alongside us to windward. Caught in her lee, we seemed to be drawn ever closer to her clifflike steel hull, and our mast, swinging wildly like an inverted pendulum as we wallowed in the confused seas, came close to hitting her side. Had it done so the mast would have snapped like a match, and we would really have needed help.

At this point a very strong squall came through, bringing with it more rain and spray that stung our faces, half blinding us. We needed to get away from the ship before there was a disaster, so we bent on the first foresail that came to hand and ran off as fast as we could, though it felt as if the rigging would give way under the strain. Luckily nothing broke and we quickly went back to bare poles. The *Janetta* moved off, disappearing quickly into the murk, leaving us feeling relieved but lonelier than ever.

As the day wore on the wind gradually eased and some clear patches began to appear among the fast-moving clouds, which were now a little higher in the sky. We hoisted some sail and resumed our proper course. We took no sights but there was a magnificent ruddy sunset that edged the black clouds with vermilion. Timed against the chronometer it placed us at about 37 degrees West.

BORN TO A well-heeled family near the southern French town of Albi in 1741, La Pérouse joined the French navy in 1756 and during his long career saw action in North America, the West Indies, and the Indian Ocean. While serving in the Isle de France (Mauritius) in the 1770s, he fell deeply in love with the beautiful young Eléonore Broudou (born in 1755), daughter of the manager of the local naval store yards and hospital. Although French-born, she was far from being a suitable match in the eyes of La Pérouse's father.[1] The devoted Eléonore followed La Pérouse when he was posted back to France in December 1776, but his father remained adamantly opposed to their marriage:

> *You make me tremble, my son! You are cold-bloodedly contemplating the consequences of a marriage that would*

> *disgrace you in the eyes of the Minister [in charge of the navy] and cause you to lose the protection of powerful friends! . . . You are preparing nothing but regrets for us; you are sacrificing your fortune and the respectability of your condition to a frivolous beauty and to so-called attractions which perhaps only exist in your imagination.*[2]

When La Pérouse returned in 1783 from service in the American Revolutionary War, he had reached the rank of post-captain and had been granted both a knighthood and a pension by the king. He was now forty-two. Eléonore continued to wait patiently for him, but his parents were determined that he should marry an aristocratic girl from Albi whom he hardly knew. La Pérouse offered poor Eléonore a large sum of money to be freed from his earlier promise to marry her. Seemingly cruel, this seems to have been a stratagem on his part, and one trusts that Eléonore was a party to it. She refused the offer but gave him his freedom, protesting for her part that she would enter a convent, as she could love no one but him. La Pérouse then rushed dramatically to meet her in Paris and wrote to his mother:

> *I was imprudent in contracting this engagement without your consent; I would be a monster if I broke my word. . . . I can only belong to Eléonore.*[3]

His parents at last relented, and in July 1783 the lovers were quietly married in Paris. A grand wedding ceremony followed at the high altar of the imposing redbrick cathedral of Albi, but they were not able to enjoy married life together for very long. While his young wife remained in Albi, La Pérouse spent much of the next two years in Paris planning the great expedition that he was soon to lead. After long preparations, and a last, fleeting visit to Eléonore, La Pérouse set sail from France in August 1785 with two well-named ships, *La Boussole* ("the compass") and *L'Astrolabe*.

The king himself took a close interest in the expedition, which was a much more elaborate operation than that which Bougainville had led. The main aim—it was frankly admitted—was to fill in, as far as possible, the gaps in the charts of the Pacific left by Captain Cook and to confirm Buache's theory about the Solomon Islands. La Pérouse was equipped with a wide variety of navigational instruments. These included: three English sextants; three astronomical quadrants; a single English "pocket watch, or chronometer, for the longitudes"; five marine clocks; and three astronomical clocks—used for checking the daily rates of the "watch" and "clocks" on dry land when opportunities occurred.[4] Two instruments for measuring variations in the earth's magnetic field ("dipping needles") that had formerly been used by Cook were lent by the Longitude Board in London, thanks to the intervention of Sir Joseph Banks, now president of the Royal Society. "I received these instruments," wrote La Pérouse, "with a feeling of religious respect for the memory of that great man."[5]

According to the official instructions—issued in the king's name—the determination of longitudes was one of the principal aims of the voyage:

> *As often as the state of the sky permits,* [La Pérouse] *will have distances between the moon and the sun or stars taken, with the instruments designed for this purpose, to determine the longitude of the vessel, and will compare the result with that indicated by the clocks and marine watches at the same place and same time: he will take care to multiply the observations of each kind, so that the mean result of the two operations may yield a more precise determination.*[6]

The scientists accompanying La Pérouse were advised to preserve their original lunar-distance calculations so that it might later be possible to correct their results in the light of any new astronomical observations made on dry land.[7] La Pérouse boasted that his officers never missed an opportunity to measure their longitude by lunars when the weather was favorable. They were so well practiced and so well supported by Joseph Lepaute Dagelet, the official astronomer aboard *La Boussole*, that—according to La Pérouse—their largest longitude error would not have exceeded half a degree.[8] Needless to say, they carried with them a copy of Mayer's tables and Maskelyne's *Nautical Almanacs* for the years 1786 to 1790.[9]

La Pérouse was also accompanied by a geologist, a naturalist, a mineralogist, a botanist, a gardener and three artists—two to paint natural history subjects, and one to record costume, landscape, and "in general all that is often impossible to describe."[10] It has even been suggested that an ambitious young artillery cadet named Napoleon Bonaparte was among those who wished to join the expedition. However, his domineering mother had already firmly ruled out a naval career for her son even though his teachers at military school said that he was "highly suitable to become a naval officer." So if—as seems quite possible—the young Napoleon had wanted to join La Pérouse, she would probably not have allowed him to do so.[11] Who knows what the consequences for the world might have been if she had not been such a formidable woman?

The orders given to La Pérouse were extraordinarily—almost absurdly—detailed and precise both as to the itinerary and the timetable.[12] The expedition was to take him around Cape Horn, to Easter Island, Hawaii, Alaska, and down the American coast to California, then across the Pacific to China, the Philippines, Taiwan, Korea, Japan, Sakhalin, and Kamchatka. In addition to making accurate charts of the places they visited, La Pérouse and his colleagues were required to gather every kind of information that might conceivably be useful—meteorological, botanical, social, anthropological, economic, political, military, and much more besides. The list is daunting and might have overawed a less determined individual. But despite the enormous care that had gone into the preparations, and the skill and experience of La Pérouse and his officers, the voyage was marked by calamities worse than any experienced by Cook and other recent explorers.

Disaster struck first in July 1786, when they were exploring the Pacific coast of present-day Alaska.[13] Pleased to have come so far without one man falling sick or even suffering an attack of scurvy, La Pérouse and his companions were anchored in a large bay (now known as Lituya Bay). Three of the ships' boats were dispatched to take soundings. La Pérouse had expressed some misgivings about the man he put in command of this operation, the first lieutenant of *La Boussole*, Charles d'Escures, chevalier de St. Louis, "whose zeal had sometimes seemed to me a little ardent." So he gave him written orders in which he specifically forbade him to go near the entrance of the bay if this presented any danger. The chevalier bristled, but La Pérouse mildly pointed out to him that young officers often took foolish risks—risks much better avoided on such an expedition as the one on which they were now engaged. Sadly, his warnings were ignored. The boats set off early in the morning, in fine weather, their crews planning to take a picnic ashore and go hunting, as well as to conduct the necessary survey work: it was to be as much a pleasure trip as duty. A few hours later the smallest of the boats returned, dismally alone. It soon emerged that the two larger boats and all their occupants had been lost in the violent race thrown up by the extremely powerful tide in the narrow entrance to the bay—which ran at "three or four leagues an hour." Worse still, contrary to standing orders, two young twin brothers had been allowed to leave the ship together. They had both drowned and, despite the help of the natives, none of the crews of the two missing boats was recovered alive. The effect of this disaster on the morale of the expedition can easily be imagined. Before departing, La Pérouse erected a memorial to the men who had lost their lives. Beneath it a bottle was buried containing the message: "At the entrance to the port, twenty-one brave sailors perished. Whoever you may be, mix your tears with ours."[14]

This catastrophe could not, of course, be allowed to disrupt the voyage. Some months later, having cruised down the coast of British Columbia and on to the Spanish colony of Monterey in California, La Pérouse set out for the Mariana Islands, on the far side of the Pacific. When they were roughly halfway and soon after passing Necker Island (a remote member of the Hawaiian archipelago that he named after a celebrated former French minister of finance, perhaps because it was so barren), they were very nearly wrecked during the night on a reef slightly farther to the west.[15] It was a very calm night, and La Pérouse had wisely given orders for the two ships to proceed slowly. His caution was well founded as it was not until they were within four hundred yards of the rocks that they spotted the breakers, which were so slight that they might easily have escaped notice until it was too late. Others would have breathed a sigh of relief and sailed on, but La Pérouse was too professional for that: "I was persuaded that if we did not reconnoitre this little rock more closely, many doubts would have remained about its existence."[16] Having earlier in the voyage wasted days trying to find a nonexistent rock reported in the South Atlantic, he was also conscious of the time and energy that others would have to devote to the rediscovery of the reef if he did not accurately determine its position. So in the morning they reversed their course and found the islet and its surrounding reefs again, Dagelet fixing its coordinates as best he could. The longitude he recorded was impressively accurate.

Now known as La Pérouse Pinnacle, the island forms part of French Frigate Shoal, so named by its discoverer because it had almost brought his vessel's career to an early end.[17]

After an uneventful passage, La Pérouse reached the Mariana Islands on December 14, 1786, commenting that "the method of distances [that is, lunars], especially when combined with the use of marine timekeepers, leaves so little to desire . . . that we made our landfall . . . with the greatest precision."[18] He continued on his way, travelling via Macao to Manila, and then up through the Sea of Japan and on to Kamchatka, making friendly contact on the way with the Ainu people of northern Japan. From the town of Avatscha (now Petropavlovsk)—a remote corner of the Russian Empire earlier visited by the *Resolution* and the *Adventure* following Cook's death—the Russian-speaking Barthélémy de Lesseps[19] undertook an epic trans-Siberian journey to bring home copies of La Pérouse's journal and charts.[20]

Having received fresh orders from Paris (and a promotion to the rank of *chef d'escadre,* or commodore)[21] La Pérouse sailed southeast, crossing a little-known and almost empty expanse of the Pacific where he made no new discoveries, before calling at the Navigator Islands (earlier visited by Bougainville, and now called Samoa) to obtain desperately needed fresh provisions. The events that unfolded here in December 1787 illustrate all too clearly the dangers that explorers faced when dealing with warlike native peoples.[22]

The island of Maouna (now known as Tutuila) was surrounded by reefs that prevented the two ships from approaching close to the shore, but their needs were so great that La Pérouse decided to risk anchoring outside them. There was a heavy onshore swell, and he was anxious that if their cables parted and the wind dropped they would be set onto the reefs—just as Cook had been. Although it was already late in the day, the commander of *L'Astrolabe*, Paul-Antoine Fleuriot, vicomte de Langle, made a brief foray ashore with a party of sailors and soldiers in three boats that could pass through the small gaps in the reef, and was apparently well received by the inhabitants.

The following morning the two ships were surrounded by canoes full of islanders who seemed keen to trade, though they cared nothing for the hatchets and cloth that had proved so popular elsewhere: they were interested only in glass beads. This time, La Pérouse and de Langle both went ashore in the ships' boats and set up camp on the beach, the perimeter guarded by a line of soldiers. Trading began, but soon got out of hand:

> *The women, some of whom were very pretty, offered their favours to anyone who had beads to give them. . . . Soon they tried to cross the line of soldiers, who resisted them too feebly to stop them; their manners were gentle, cheerful and seductive. Europeans who have gone round the world, above all Frenchmen, have no weapons in the face of such an attack.*[23]

The eager women quickly breached the ranks of soldiers, and confusion reigned. Order was eventually restored, but La Pérouse suspected that the local men—who were tall and strongly built—had underestimated the power of European firearms. He therefore arranged for three pigeons to be freed and shot in full view of everyone. This seemed to have the required effect, and while the water casks were being filled, La Pérouse wandered over to the neighboring village, where he admired the sophisticated design of a house that he supposed belonged to a chief:

> *The best architect could not have given a more elegant curve to the ellipse that formed the ends of this hut: a row of columns . . . surrounded it: these columns were made out of very nicely worked tree trunks, between which finely*

> *made blinds covered in fish scales could be raised or lowered with cords.*[24]

Back on the beach, trading had been brisk and all seemed to be going well. Like Bougainville in Tahiti, La Pérouse and his companions were ravished by the beauty and abundance of the island:

> *These islanders, we kept saying, are without doubt the luckiest people in the world: surrounded by their women and their children . . . they have no other care than to raise birds and, like the first man, to gather without effort the fruits that grow above their heads.*[25]

However, all was not as innocent as it seemed. They saw no sign of weapons, but could not help noticing that the male islanders were covered in scars and looked ferocious. With the benefit of hindsight, La Pérouse caustically commented that "nature had doubtless left this imprint on the bodies of these Indians" as a warning that "the almost savage man, living in a state of anarchy, is a more dangerous being than the fiercest of animals."[26]

Having returned to the ships, La Pérouse was keen to set sail, but de Langle had found an ideal place to fill more water casks and was convinced that the members of the crew who were suffering from scurvy would benefit from a run ashore. Although La Pérouse was concerned that the two frigates were too far offshore to provide any protection to a shore party, de Langle persuaded him that another expedition should land the next day.

The two frigates weighed anchor—just in time, as it turned out that the cable of *La Boussole* had almost parted—and stood off and on under sail throughout the night. At 11 A.M. they lay a few miles from the island. La Pérouse dispatched several boats to the new watering place, containing all the victims of scurvy together with a large number of armed men—sixty-one in all. But to the dismay of all concerned, it turned out that the supposedly ideal landing place could now be reached only by a narrow, tortuous channel beyond a bar on which the swell was breaking: it was approaching low water and de Langle, who had first seen the bay when the tide was high, had underestimated the rise and fall. While La Pérouse remained helplessly aboard *La Boussole*, de Langle proceeded to the shore and began filling the water casks. By mid-afternoon a huge and menacing crowd of natives had gathered around de Langle and his men, and he decided to embark. But the tide had now ebbed further and the bay was almost dry, leaving the two larger boats hard aground:

> *If the fear of starting hostilities and of being accused of barbarity had not constrained M. de Langle, he would doubtless have ordered a discharge of muskets . . . which would certainly have driven off this multitude; but he flattered himself that he could contain them without spilling blood and fell victim to his own humanity.*[27]

De Langle attempted to pacify the enormous crowd by offering gifts to some whom he took to be chieftains, but this gesture served only to provoke the others. When the attack came, it was sudden and furious: de Langle was the first to fall under a fierce rain of stones and was soon clubbed to death. Many of the firearms had been soaked by seawater and were useless, and although some men escaped to the smaller boats that remained afloat, no one remained alive on the larger ones. Out of sixty-one men, twelve died and twenty were wounded.

When the survivors reached the ships, many canoes were milling around them with produce to sell. La Pérouse had great difficulty in containing his own rage and that of his crew, but he succeeded in preventing reprisals.[28] Realizing that it would be impossible to recover the bodies of his dead comrades, or indeed the two larger boats, he reluctantly decided to quit the island, and headed west for Botany Bay, which he reached in January 1788 via Tonga and Norfolk Island. La Pérouse was, not surprisingly, deeply dejected. He wrote to a friend:

> *Whatever professional advantages this expedition may have brought me, you can be certain that few would want them at such a cost, and the fatigues of such a voyage cannot be put into words. When I return you will take me for a centenarian, I have no teeth and no hair left and I think it will not be long before I become senile. . . .*[29]

After refreshing themselves, and sending home another batch of papers in a British ship, La Pérouse and his remaining comrades set sail again on March 10, with the aim of returning to France in the summer of 1789. First, however, they planned to pin down the location of the long-lost Solomon Islands. Neither *La Boussole* nor *L'Astrolabe* was ever seen again.

In France it eventually became clear that La Pérouse, with his two ships and the two-hundred-odd members of their crews, had been lost. Their disappearance caused consternation and was a mystery that demanded to be solved. By now the French Revolution was gathering momentum, but thanks in part to the king's intervention, an elaborate search-and-rescue mission was mounted under the command of Rear Admiral Joseph-Antoine Bruny d'Entrecasteaux. Two ships—touchingly renamed *La Recherche* ("search") and *L'Espérance* ("hope")—sailed from Brest in September 1791, but the voyage was anything but prosperous. Disease and scurvy cut through their crews, and d'Entrecasteaux, along with his senior colleague Jean-Michel Huon de Kermadec, died far from home. Although they made important discoveries on their way, and confirmed at last the existence of the Solomon Islands, they found no trace of the missing ships and their crews. The king's interest in the fate of La Pérouse did not wane even though his own survival was by now in grave doubt. Shortly before his execution in 1792 he is said to have asked: "Is there any news of Monsieur de La Pérouse?"[30]

The publication in 1797 of the account of the voyage of *La Boussole* and *L'Astrolabe*, based on the materials La Pérouse had sent home before his disappearance, turned him into a national hero, and rumors circulated about the possible survival of some of the crew. Plays and musicals about the expedition were performed all over Europe,[31] but it was not until 1827 that an intriguing adventurer named Peter Dillon managed to identify the small, remote, and pestilential island of Vanikoro in the Santa Cruz group—so named by Mendaña and Quirós in 1595—as the place where La Pérouse's voyage had come to its abrupt and disastrous end. Ironically, d'Entrecasteaux had sailed quite close to it—without stopping—as had the *Pandora* when searching for the *Bounty* mutineers. The *Pandora* even saw smoke rising—perhaps from a signal fire lit by the survivors—but reasoning that mutineers would not wish to draw attention to themselves, they did not bother to investigate.

Dillon was a colorful character who might have stepped out of the pages of a Conrad novel. Brought up by Irish parents on the island of Mauritius, he spoke French as well as English, and by the time he solved the mystery of La Pérouse's fate he had been knocking around the South Seas for some years, trading among the islands and—by his own account—narrowly escaping death at the hands of cannibals in Fiji. In 1826 he picked up the trail of La Pérouse on the island of Tukopia in the Santa Cruz group, where he came across various artifacts of European origin, including a silver sword-guard, said to come from the nearby island of Mannicolo (now Vanikoro).[32] Natives also reported that two large ships, like the one in which Dillon was sailing, had been wrecked there many years earlier. Having tried unsuccessfully to land on Vanikoro, which was surrounded by dangerous reefs, Dillon sailed to India, where he persuaded the governor of Bengal to put him in charge of an expedition to search for any survivors from the French expedition and to gather evidence of what had happened to the two ships and their crews.[33]

After many vicissitudes, which included a short spell in prison in the town of Hobart, Tasmania, and a visit to New Zealand, Dillon at last reached Vanikoro in September 1827. Despite language difficulties, he was able to piece together testimony from some of the older local natives about the loss of the two ships. Both had apparently been wrecked on the reefs that surround the island in a hurricane. Some of their crews were said to have reached the shore in safety, despite attacks by sharks, only to be killed by natives who supposed them to be ghosts.[34] Dillon was led to believe that some survivors had built a small vessel in which they had departed, leaving others behind.[35] There were even suggestions that a few might still be living on neighboring islands, though ill health prevented Dillon from extending his cruise long enough to confirm this.[36]

On Vanikoro he acquired by barter a large and miscellaneous haul of material that had been recovered from the wrecks by the natives, including a brass bell inscribed "Bazin m'a fait"[37] ("Bazin made me"), a silver candlestick with a coat of arms, part of a European ship's stern, and a grindstone. His men also managed to recover some small brass guns from one of the wreck sites.[38] On Dillon's eventual return to Europe he travelled to Paris, where the sixty-four-year-old de Lesseps was able to identify many of the relics.[39] The coat of arms was variously attributed to Collignon, a botanist aboard *La Boussole*, or to de Langle, the commander of *L'Astrolabe*.[40] Dillon was presented to King

Charles X and made a chevalier of the Légion d'Honneur, as well as being awarded an annuity of four thousand francs in recognition of his services.[41]

In February 1828, a Frenchman, Jules Dumont d'Urville, sailed to Vanikoro and located what he believed to be the site of the wreck of *L'Astrolabe*. He found more objects on the seabed, but sickness among his crew forced him to leave without obtaining "irrefutable" proof of his discovery. An expedition in 1883 recovered anchors, which now lie at the foot of a monument to La Pérouse in Albi.[42] In much more recent times a series of well-equipped expeditions have shed new light on the fate of La Pérouse and his companions. The wreck sites of both ships have been identified and, on shore, the remains of the camp used by the survivors have been discovered. It is still not certain that any of the crew escaped from Vanikoro, though according to local traditions, descendants of one of the ship's surgeons are living to this day on a neighboring island. Many more artifacts have been recovered from the wrecks, among which the most significant (found in 2005) is a sextant, inscribed by its maker "Mercier, Brest." Just such an instrument was recorded in the inventory of *La Boussole*—handled, perhaps, by La Pérouse himself.

La Pérouse's widow, Eléonore, struggled for years to obtain the pay that was due to her late husband. She was promised a share in the proceeds of the magnificent three-volume account of the voyage edited by the Baron Milet-Mureau and published in 1797. However, it sold poorly. In 1803 Napoleon granted Eléonore a pension and a rent-free royal apartment at the chateau of Vincennes, though she chose not to live there.[43] She died in 1807, aged fifty-two, long before the mystery of her husband's disappearance had been solved.

Chapter 11

The Travails of George Vancouver

Day 12: At last the weather is improving and so is our mood. Skies almost clear, wind moderating and a warm sun—clothes, oilskins, boots and sleeping bags all drying.

Took sun sight at 0915 and raised main. Now being headed by a NE wind and it's getting a bit light. Changed up to the blue genoa. Had our first contact with Europe when we picked up a Portuguese radio station, presumably in the Azores, which are now less than five hundred miles away. Noon position: 43°02.5' N, 35°23' W—by earlier sun sight crossed with mer alt. Several runners on the mainsail bolt rope had come adrift: Alexa and I reattached them. We also sewed up the tear in the sprayhood. Skipper did some maintenance on the self-steering gear. Shearwaters and petrels still around us.

Saw dozens of dolphins flying through the waves—they move at an amazing speed with enormous grace. That brought smiles to our faces!

Weather still fair but wind now right on the nose. Some cirrus in the west suggests another depression may be coming but barometer steady.

Noticed a tear in the big genoa and replaced it with heavier white one which we haven't used before.

Heard BBC shipping forecast for first time. Fine evening. Colin and I did some star sights—Mirfak, Altair and Arcturus gave us good fix consistent with noon position.

During the 1790s British sailors were prominent among those who continued to fill in the gaps in the charts of the Pacific. None achieved more than George Vancouver, though the portly, unheroic gentleman revealed in the portrait that is our only record of his appearance gives no hint of the reserves of courage and determination on which he must have drawn. While the Canadian city and the great island that bear his name keep his memory alive, he is hardly known in his own country.

Vancouver was born in 1757 in King's Lynn, then a busy port on the east coast of England, in the county of Norfolk, where his father was a customs officer. Not much is known about George's childhood or his antecedents, though his name presumably has a Dutch origin. In 1772, possibly through the influence of the eminent musician Charles Burney, a former neighbor who had served as organist at the parish church of King's Lynn, George obtained a place aboard the *Resolution*.[1] He was fourteen. On Cook's third voyage Vancouver served as a midshipman on the *Discovery*, commanded by Charles Clerke.[2] He idolized Cook and plainly learned his trade well: on his return home in 1780 he was promoted to lieutenant and in 1782 he was posted to the West Indies station, where he was made fourth officer aboard the *Fame*, a seventy-four-gun ship.[3] Vancouver soon found himself at home again on half pay, but in 1785 he was back in the West Indies as third lieutenant of the *Europa*, the flagship.[4] While serving there, Vancouver surveyed the harbors of Kingston and Port Royal, Jamaica, and by the time he was posted home in 1789 he had risen to first lieutenant of the *Europa*.

In 1790 Vancouver was appointed first lieutenant of the *Discovery*, a newly built 337-ton sloop that was being fitted out for a voyage of exploration in the Pacific.[5] This plan was shelved as a result of the so-called Spanish Armament—a brief but alarming international crisis that followed the seizure by the Spanish of British trading vessels and shore facilities at Nootka Sound, a harbor on the west coast of what is now Vancouver Island that Cook had earlier visited. The Spanish—who had unrealistically laid claim to the entire west coast of North America—quickly backed down in the face of strong British protests, but by then the original commander of the *Discovery* had been assigned to other duties. Vancouver now had his chance. In November 1790 he was given the command of the *Discovery* and was ordered to sail for the northwest coast of America to explore the region between latitudes 60 and 30 degrees North.[6] This was a challenge for which he was well qualified, but the voyage also had another purpose: to negotiate the return of the British property seized by the Spanish at Nootka Sound. In this delicate task Vancouver was not successful—probably because of the inadequacy of his instructions. He was not the first diplomat to feel let down by his superiors.

In the introduction to the account of his own great voyage, Vancouver described how "the commercial part of the British nation" responded to Cook's discoveries in the Pacific—particularly by developing a trade in sea-otter furs from the northwest coast of America, which fetched enormous prices in China. His scorn for the greed and amateurism of these traders is matched by his disdain for the armchair geographers who dared to criticize his mentor:

> *Unprovided as these adventurers were with proper astronomical and nautical instruments, and having their views directed almost intirely to the object of their employers, they had neither the means, nor the leisure, that were indispensably requisite for amassing any certain geographical information. This became evident, from the accounts of their several voyages given to the public; in which, notwithstanding that they positively contradicted each other . . . they yet agreed in filling up the blanks in*

> *the charts of Captain Cook with extensive islands, and a coast apparently much broken by numberless inlets, which they had left almost intirely unexplored.*[7]

Vancouver complained that the unreliable charts accompanying the accounts of their voyages had "roused from slumber" the "favourite opinion that had slept since the publication of Captain Cook's last voyage" that a northeastern passage existed between the waters of the Pacific and North Atlantic oceans.

Bougainville had, in his day, complained about the "lazy, proud writers who, philosophizing in the shade of their studies . . . imperiously submit nature to their imaginations."[8] Such speculators had then insisted on the existence of the great southern continent, but now it was up to Vancouver to vindicate Cook by proving that the supposed passage through the North American landmass was also a fantasy. However, he also had scientific ambitions of his own:

> *Among other objects demanding my attention, whilst engaged in carrying these orders into execution, no opportunity was neglected to remove, as far as I was capable, all such errors as had crept into the science of navigation, and to establish, in their place, such facts as would tend to facilitate the grand object of finding the longitude at sea; which now seems to be brought nearly to a certainty, by pursuing the lunar method, assisted by a good chronometer.*[9]

Vancouver spent three summers (1792, 1793, and 1794) exploring and charting the coast from southern California to Alaska, retiring each autumn to the Hawaiian archipelago—large parts of which he also charted—when bad weather made survey work on the American shore impossibly difficult. Cook had begun the

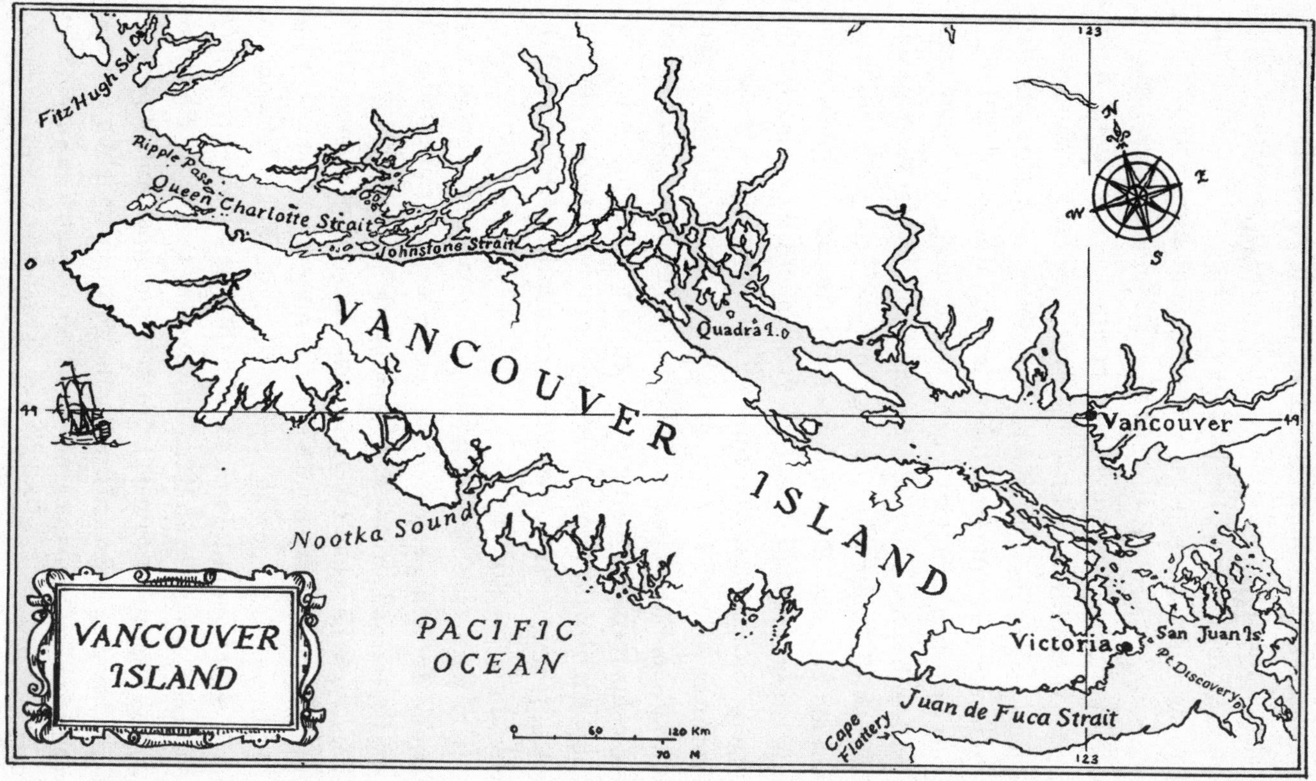

Fitz Hugh Sd.
Ripple Pass
Queen Charlotte Strait
Johnstone Strait
VANCOUVER ISLAND
Quadra I.
123
N
E
W
S
49
Vancouver
Nootka Sound
VANCOUVER ISLAND
PACIFIC OCEAN
Victoria
San Juan Is.
Pt. Discovery
Juan de Fuca Strait
Cape Flattery
0 60 120 Km
70 M

task, but a huge amount of work remained. As Vancouver soon realized, the extraordinarily intricate network of islands, channels, and inlets presented an even greater surveying challenge than he had imagined. He showed astonishing patience and dedication as a surveyor, but he lacked Cook's natural leadership qualities. Ill-tempered and erratic, he was a tough disciplinarian, and his prickly self-importance often made him an object of ridicule. Allowance must, however, be made for the fact that he was suffering from the pulmonary illness that was to cut short his life not long after he returned home, and he also carried another heavy burden. Unlike Cook, Vancouver was lumbered with a gang of teenage grandees whose well-connected parents had used their influence to have them shipped off on this latest voyage of discovery. Presumably, they regarded the expedition as a character-building alternative to the traditional continental Grand Tour, which was now precluded by the French Revolution. But, as Jonathan Raban puts it, these "patrician adolescents regarded their captain with a mixture of raw fear and snobbish disdain."[10] Vancouver's biographer has described the midshipmen as varying "between the totally ineffective and the potentially rebellious."[11]

The worst was the Honorable Thomas Pitt, son of Lord Camelford, and a cousin of the prime minister of the day. He was sixteen and already had a bad reputation: his previous commanding officer had taken the unusual step of refusing to give him a signed certificate of service, commenting laconically that "during the time he was under my Command his Conduct was such as not to entitle him to it."[12] During the outward voyage, the troublesome Pitt fell foul of Vancouver, who twice had him tied to a gun and caned in front of the other midshipmen—a punishment known as "kissing the gunner's daughter."[13] Later, when he was found to have been asleep on watch—a very serious breach of discipline—Vancouver had him put in irons in front of the whole crew.[14] These indignities Pitt neither forgave nor forgot, and though this obnoxious young man probably deserved his punishments, he made a bad enemy. Eventually dismissed and sent home to England, Pitt was to make the last years of Vancouver's life a misery.

As La Pérouse's comrades discovered to their cost, the tidal streams of the Pacific Northwest seaboard are in some places exceptionally strong, occasionally running at speeds of well over ten knots. Square-rigged sailing ships with no auxiliary power would have been completely unmanageable in currents even half that strong and would have been in great danger if it was too deep to anchor—as was frequently the case.

Today, cruise ships ply up and down this coast, but the many uninhabited islands cannot have changed much since Vancouver's time, and there are still a few small areas that remain uncharted. Dark coniferous forests that have never been cut enfold the hills and mountains, advancing almost to the water's edge, and there is even more rainfall than in England. Until I first went ashore on an island in a remote corner of northern British Columbia, I had no idea what the words "impenetrable forest" really meant: the dripping trees were so tightly packed that it was impossible to force a passage between them. Short of using an axe, the only practical way of getting beyond the beach is to follow a stream or river. No wonder the native tribes lived entirely on the shoreline. At anchor, by night, surrounded by high mountains, far from any human settlement, with the rain falling steadily, the silence and stillness are broken only by the occasional call of an owl or wolf. To the modern eye, this country has the Romantic beauty of an untouched wilderness, and so too did it strike some of the "young gentlemen" aboard *Discovery*, but to Vancouver the solitude and gloom were simply oppressive. As his health declined he must have looked forward eagerly to their spells in the warm waters of the Hawaiian Islands.

Vancouver set sail from England on April 1, 1791 (the significance of this date was not lost on his crew) with two ships, the *Discovery* and the much smaller brig, the *Chatham*. The *Discovery* was Vancouver's first command: it soon became clear that, though new, she had not been well built. Having called at the Cape of Good Hope to refresh their supplies, rate the chronometers, and undertake essential repairs, the two small ships crossed the Indian and Pacific oceans. After making stops in Australia, New Zealand, Tahiti, and Hawaii, they finally made their landfall on the coast of North America on April 17, 1792. Vancouver's account illustrates the challenges then facing a navigator approaching a poorly charted coast after a long ocean passage.

A month out from Hawaii, a single set of lunars taken on April 15 suggested that the *Discovery* was 232°56½' East of Greenwich; the chronometer, however, indicated a figure of 232°7¾', while DR gave 229°39': a maximum difference of almost 160 nautical miles. The latitude was 37°55'—not far north of San Francisco. Vancouver knew that he must be approaching land, but the following day the wind increased and it looked as if a storm was on the way. No soundings were to be had with the 120-fathom line, and as Vancouver could not safely count on the charted longitude of the coast they were approaching, he stood off until daybreak, when they resumed their northeasterly course in a rising gale. In the afternoon, the *Discovery* was forced to head south under storm canvas and lost touch with the *Chatham* during the night. Having renewed contact with her, they headed once more for the land. "The sky being tolerably clear," Vancouver took six sets of lunar observations that gave a longitude 50 minutes to the eastward of the chronometer:

> *Soon after mid-day we passed considerable quantities of drift wood, grass, sea weed, &c. Many shags, ducks, puf-*

fins, and other aquatic birds were flying about; and the colour of the water announced our approach to soundings. These circumstances indicated land not far off, although we were prevented seeing any objects more than 3 or 4 miles distant, by the weather, which had become very thick and rainy. Being anxious to get sight of the land before night if possible, we stood to the eastward with as much sail as we could carry, and at four in the afternoon reached soundings at the depth of 53 fathoms, soft brown sandy bottom. The land was now discovered . . . at the distance of about 2 leagues, on which the surf broke with great violence. . . .

During the night Vancouver plied up and down under an "easy sail" in order to be near the land in the morning, but it remained invisible until a light breeze revealed the shore to the northeast:

The observed latitude was at this time 39°27'; the longitude 235°41'30"; by the chronometer 235°. The former was deduced by the mean results of eighty-five lunar distances. . . . This made the chronometer 41'30" to the west of that which I supposed to be nearest the true longitude;† and from the general result of these observations it evidently appeared, that the chronometer had materially altered in its rate since we had reached these northern regions.*[15]

Vancouver had learned his navigational skills from the astronomer William Wales aboard the *Resolution*. From Cook, too, he must have learned much about survey techniques. Although in

* The text says 45" but this is clearly a mistake as the mean result of the lunar observations is shown as 30".

† In this latitude that is a distance of almost thirty nautical miles.

the published account of his own voyage, Vancouver (like Cook before him) is discreet about the exact methods he employed, we know that in 1788 he had surveyed the harbor of Kingston, Jamaica, by laying out an accurately measured baseline and then fixing the position of every landmark and shoal "by intersecting Angles taken by Sextant & protracted on the spot."[16]

On board *Discovery*, Vancouver's equipment included an astronomical quadrant for use whenever a temporary observatory was set up on land, and no fewer than twelve sextants. At this point the two ships shared just two chronometers between them—one by Kendall ("K3" aboard *Discovery*) and one by Arnold (aboard *Chatham*). K3, which had just been overhauled by its maker, had accompanied Cook on his last voyage and again proved excellent. Vancouver was very pleased to have it. A store ship, which was to join Vancouver later that summer in Nootka Sound, was to bring out the official astronomer, William Gooch, as well as a great deal more equipment, including an astronomical regulator clock, more sextants and chronometers, a pocket watch with a second hand, theodolites, and a measuring chain.[17]*

Having made his landfall, Vancouver sailed north, conducting a "running survey" on the way, just as his hero Cook had done on so many occasions, using horizontal sextant angles and compass bearings to establish the relationships between the features observed. Sketches were made to record the views. A north-south–trending coast, like the one Vancouver was now exploring, helpfully allowed him to check his progress by means of straightforward latitude observations. So long as the winds were favorable, such a survey could proceed quite quickly, but if the ship was working to windward she would be obliged to follow

---

* The pocket watch, however, was broken by one of the ship's cats at the very start of the voyage. Gooch made excuses for her: "she is a very young cat & perhaps its beating attracted her notice."

a zigzag course. This would not only be slower, but would also greatly complicate the task of the surveyor.

Having passed the entrance of the great Columbia River without catching sight of it, Vancouver reached Cape Flattery and entered the Strait of Juan de Fuca—a wide inlet that Cook had failed to notice—and his detailed survey work began. The extraordinary complexity of the task soon became apparent as they discovered dozens of islands to the north and numerous fjords disappearing into the high, snowcapped mountains to the east, all of which had to be investigated. On June 22, 1792, Vancouver was startled and disconcerted to encounter two small Spanish naval vessels also engaged in survey work, but he immediately established friendly relations with their commanding officers, Galiano and Valdés,* and they agreed to collaborate.

It was impossible to conduct such a complex survey from the ships, especially in dangerously tidal waters. Having found a secure anchorage, the ships' boats (large rowing boats, though they could also carry sails) were sent ahead to take horizontal sextant (or, better still, theodolite) angles from locations on dry land that commanded extensive views whenever opportunities offered. Vertical sextant angles would have been helpful for determining the heights of mountains. The irregular terrain would often have precluded the physical measurement of baselines, but a useful makeshift was the use of sound: by timing the interval between the puff of smoke from a gun and the arrival of the report in still air, a good estimate could be made of distance. Noon observations for latitude would be taken with the sextant

---

* Dionisio Alcalá Galiano and Cayetano Valdés y Flores had been detached from Alessandro Malaspina's great expedition. Malaspina was a Neapolitan subject of the Spanish king whose ambitious and well-equipped voyage of discovery from 1789 to 1794 doubled as an inspection of the Spanish colonial empire. His liberal recommendations were badly received after his return, and he ended up in jail. Galiano was to die in 1805 fighting the British at the Battle of Trafalgar.

using an artificial horizon* if the actual horizon was not visible. Tidal measurements also had to be made—of both heights and intervals, and their relationships to the passage of the moon.

Such expeditions would sometimes last several weeks. This was exhausting and occasionally hazardous work for the crews, as the boats often had to be rowed over long distances, food supplies sometimes ran low, and the natives were not always pleased to see the European visitors. While the ships were at anchor, an observatory would be set up in a tent on shore to check the longitude by lunars (sometimes hundreds of sets were taken over many days) and to rate the chronometers. When all the boats had returned, the two ships would gingerly move forward to a new anchorage and the process would start afresh.

Each observatory site was meant to provide a more or less precise geographical location to which all the other survey measurements could be reduced. However, while Vancouver's latitudes were usually very accurate, the longitudes he recorded in these northerly regions often differed significantly from modern values. During the first season the longitudes he established by lunars at his three observatories were all too far to the east, in the first case by an average of 14.8 minutes—about 9.5 nautical miles in these latitudes. Vancouver was more successful in the second season (by then he had five chronometers at his disposal), but even then his longitudes still tended to be too far to the east.

A modern analysis reveals that the shortcomings in the first season's surveys were largely due to inaccuracies in the predictions contained in the *Nautical Almanac*, especially those relat-

---

* A small bath of mercury in which the angle between the reflected image of the sun and the sun itself could be measured: the correct observed altitude being half the resultant figure.

ing to the positions of the moon.[18] Had the *Almanac* not been in error, the crucial position Vancouver assigned to his first observatory at Port Discovery would have been out by only a few hundred yards. Vancouver plainly had difficulty in reconciling the inconsistencies between the results of his lunar observations and the chronometer readings, but chose to rely on the former—in keeping with official guidance. His navigational task was also complicated by other factors. Poor weather conditions often made it impossible to obtain sights for weeks on end (during the last season he was unable to make any lunar-distance observations), and his deteriorating health was also a serious handicap. In the later stages of the voyage Vancouver was frequently confined to his cabin and was unable to supervise the astronomical observations personally.[19]

Sailing in company with the Spanish vessels, Vancouver headed in a northwesterly direction toward the intricate maze of islands that separate the northeastern shores of Vancouver Island from the mainland. This was often nerve-racking work. As night fell they entered a "spacious sound stretching to the eastward" where Vancouver wished to stay until daybreak, but it was too deep to anchor even close to the shore:

> *The night was dark and rainy, and the winds so light and variable, that by the influence of the tides we were driven about as it were blindfolded in this labyrinth, until towards midnight, when we were happily conducted to the north side of an island. . . . At break of day we found ourselves about half a mile from the shores of a high rocky island, surrounded by a detached and broken country, whose appearance was very inhospitable. Stupendous rocky mountains rising almost perpendicularly from the sea, principally composed the north west, north and eastern quarters. . . . The infinitely divided appearance of the*

region into which we had now arrived, promised to furnish ample employment for our boats.[20]

Having gone ashore to fix their position, Vancouver and his colleagues moved on, but the next anchorage was no more cheerful:

> *Our residence here was truly forlorn: an aweful silence pervaded the gloomy forests, whilst animated nature seemed to have deserted the neighbouring country, whose soil afforded only a few small onions, some samphire, and here and there bushes bearing a scanty crop of indifferent berries. Nor was the sea more favourable to our wants, the steep rocky shores prevented the use of the seine* [net], *and not a fish at the bottom could be tempted to take the hook.*[21]

There were unexpected hazards, too. One of the boats discovered a deserted native village and, while examining the abandoned dwellings,

> *our gentlemen were suddenly assailed by an unexpected numerous enemy, whose legions made so furious attack upon each of their persons, that unable to vanquish their foes, or to sustain the conflict, they rushed up to their necks in water. This expedient, however, proved ineffectual: nor was it until after all their clothes were boiled, that they were disengaged from an immense hord of fleas, which they had disturbed by examining too minutely the filthy garments and apparel of the late inhabitants.*

Needless to say, none of the various inlets that Vancouver and his Spanish colleagues investigated on the mainland proved to be the entrance to the fabled passage to the Atlantic, but they did find a channel leading to the northwest that eventually connected to the open sea—the Johnstone Strait, famous now as the home of a large population of orcas. It was now clear that the land to the south was an island. Despite his immense experience and natural caution, Vancouver could not possibly navigate these waters without risk, and in July 1792, after parting company with the slower-moving Spanish vessels, the *Discovery* nearly came to grief in Ripple Passage. Having picked her way through the islands on the mainland shore, she had almost reached the open sea when the light southwesterly breeze dropped completely and a very thick fog descended, obscuring everything. The water being too deep for anchoring, they were left to the mercy of the currents. As Vancouver daintily puts it, their predicament "could not fail to occasion the most anxious solicitude":

> *The fog had no sooner dispersed, than we found ourselves in the channel for which I had intended to steer, interspersed with numerous rocky islets and rocks. . . . The dispersion of the fog was attended by a light breeze from the N.N.W., and as we stood to windward, we suddenly grounded on a bed of sunken rocks about four in the afternoon. . . . The stream anchor was carried out, and an attempt was made to heave the ship off, but to no effect. The tide fell very rapidly. . . . On heaving, the anchor came home, so that we had no resource left but that of getting down our topmasts, yards, &c. &c. shoaring up the vessel with spars and spare topmasts, and lightening her as much as possible, by starting the water, throwing overboard our fuel and part of the ballast. . . .*

Soon after the ship had run aground, the tide caught her stern—which was still afloat—and swung her around, at the same time making her heel so much to starboard that "her situation, for a few seconds, was alarming in the highest degree." They shored her up with timber as quickly as they could, but by the time it was low water, the starboard main chains were within three inches of the surface of the sea. Luckily the sea was entirely calm, although they were now close to the open ocean.

> *In this melancholy situation, we remained, expecting relief from the returning flood, which to our inexpressible joy was at length announced by the floating of the shoars, a happy indication of the ship righting. Our exertions to lighten her were, however, unabated, until about two in the morning; when the ship being nearly upright, we hove on the stern cable, and, without any particular efforts, or much strain, had the undescribable satisfaction of feeling her again afloat, without having received the least apparent injury.*[22]

Very shortly afterward, as the two ships continued to thread their way through a narrow, rock-strewn passage that became increasingly intricate, it was the *Chatham*'s turn to run aground. After a very anxious night, the *Chatham* eventually floated, showing little sign of damage, but the expedition had been very lucky: both ships had nearly been lost. Vancouver nevertheless pushed farther north into Fitz Hugh Sound, encountering a British vessel that brought terrible news from Nootka Sound: the commander of the store ship *Daedalus*—a personal friend of Vancouver's—and the astronomer Gooch, as well as another crew member, had all been murdered by natives at Oahu in the Hawaiian Islands.

In the light of this grim development, Vancouver cut short his northerly explorations and headed south, reaching Nootka on the seaward shore of Vancouver Island at the end of August. Here he soon established warm relations with the Spanish governor, the splendidly named Don Juan Francisco de la Bodega y

Quadra. Negotiations over the status of the Nootka settlement were conducted amicably, though inconclusively, and the governor asked Vancouver to name "some port or island" after them both "to commemorate our meeting and the very friendly intercourse that had taken place and subsisted between us." Vancouver suggested that the island on which they had first met was the perfect candidate and so they agreed that it should be called "the island of Quadra and Vancouver." As Spanish influence declined, Quadra's name was later quietly dropped, though it remains attached to a small nearby island.

While at Nootka, Vancouver heard from the master of an American trading vessel (the *Columbia*) that a large river reached the sea somewhere south of Cape Flattery, and decided to inspect it himself while sailing south to extend his survey of the Californian coast. The entrance to the Columbia River is a very dangerous place, protected by extensive shoals on which the Pacific seas break heavily. Evening was drawing in as the *Discovery* and the *Chatham* warily made their approach with a following breeze. A strong ebb tide setting out to sea gave them an easy means of retreat if the need arose, and the *Chatham*, having a shallower draft, was in the lead under the command of William Broughton. As they drew near to an apparently continuous line of breakers the soundings diminished alarmingly, and Vancouver prudently decided to turn back and spent a very uncomfortable night at anchor in the confused seas off the entrance. Broughton, however, pressed on through the white water in the gathering darkness, losing one of the ship's boats in the process. It was not immediately clear to Vancouver that the *Chatham* was safe, and he had to wait until the following morning before he saw her riding at anchor, inside the line of breakers. Vancouver's further efforts to bring the *Discovery* into the river were in vain, and he concluded that it was inaccessible to ships of more than four hundred tons burthen. Broughton sailed upriver, soon being forced by the shoaling water to take to an open boat, and travelled a hundred miles or so upstream before returning to his ship. Although the full extent of the river remained unknown, it was now clear that it did not offer a practicable route for commercial traffic.[23]

Vancouver's laborious survey work continued until the summer of 1794, by which time he had, in accordance with his orders, charted the whole American Pacific coast from 30 to 60 degrees North, thereby finally eliminating the possibility that a passage to the Atlantic through the North American continent might somewhere exist. He had also surveyed the main islands of the Hawaiian group, which—as he shrewdly recognized—occupied what was now a key strategic and commercial location. The warm relationship that Vancouver developed with the most powerful Hawaiian chief resulted in the ceding of these islands to Great Britain, though Vancouver received few thanks for this diplomatic coup on his return home in October 1795, and little advantage was taken of it.[24]

Vancouver sailed over 65,000 miles in four and a half years, and the ships' boats covered 10,000 miles—often under oars. Only five men died, and none of disease, though Vancouver's own health had steadily worsened. By the time he turned for home he had probably spent more time in the Pacific than any other European,[25] but his country was now at war with France, and although Vancouver was promoted to post-captain, his extraordinary achievements received no other public recognition. Physically unable to take another command, he had to make do with half pay and soon found himself in financial difficulties. Sharp disagreements with the troublesome botanist who had sailed with him—appointed by Sir Joseph Banks—also put the luckless Vancouver on the wrong side of Banks, a man conscious of his exalted status as president of the Royal Society.

Vancouver's greatest trials, however, arose from the malice of

Thomas Pitt, who had now inherited his father's title and great wealth. While laboriously writing the account of his long voyage, Vancouver was surprised to receive a letter from the new Lord Camelford challenging him to a duel. He quite properly responded by saying that the captain of a ship was not "called upon in a private capacity to answer for his Public conduct in the exercise of his official duty"—a judgment supported by, among others, Lord Grenville, the foreign secretary, who happened also to be Pitt's brother-in-law. Pitt then arrived in person at the Star and Garter Hotel on Richmond Hill, where Vancouver was living, and berated him. When Vancouver suggested that the propriety of his actions be tested by a senior naval officer, Pitt threatened to "drive him from the Service . . . compel him to resign his commission, and . . . wherever he should meet him, box it out and try which was the better man." True to his word, Pitt attacked Vancouver with his cane on a chance encounter in a London street, an episode that attracted the attention of the press. The famous cartoonist James Gillray published a scurrilous print making fun of Vancouver entitled "The Caneing in Conduit Street," and Banks too weighed in, complaining that Pitt and the other midshipmen had been badly treated by their captain. The whole vendetta might almost have been ridiculous had it not been so cruelly unjust.[26]

Pitt's subsequent career was, to say the least, inglorious. After murdering a fellow officer in the West Indies—an action for which he suffered no punishment—he returned to England and in 1798 was arrested while trying to reach France in a hired boat with a letter of introduction to a leading member of the French government. Since the two countries were at war, this was a capital offense, but Pitt was pardoned after appealing to King George III. In 1804, however, his luck ran out when he challenged an army officer to a duel over a "strumpet." Knowing him to be a good shot, Pitt—outrageous to the last—fired before the signal was given. He missed, and was then fatally wounded by his opponent's bullet.[27]

In steadily failing health, Vancouver struggled to finish his book and had almost done so when he died in May 1798, a month short of his forty-first birthday. His brother saw the three volumes through the press, but though reprinted they did not cause much of a stir, and Vancouver's extraordinary career and achievements were soon largely forgotten, overshadowed by those of Cook. Vancouver is buried in Petersham churchyard at the foot of Richmond Hill in southwest London. The eventual restoration of his grave with its plain headstone resulted not from the zeal of his fellow countrymen, but from that of the citizens of Vancouver. A gilded statue of Vancouver adorns the Capitol Building in Victoria, British Columbia, commemorating the man who first surveyed the province's long and rugged coast.

Chapter 12

Flinders—Coasting Australia

Day 13: Wind shifted overnight to S by W giving us good speed on 080° at last. Colin accidentally woke me an hour early at 0300 but a beautiful golden dawn made up for it. Sunrise longitude by chronometer—34° W. More cloud building up in the west and barometer falling—it looks as if we may be in for more heavy weather.

Passed by a white cargo ship with a big Fyffes banana logo sailing on the same course at 0700—the Jamaica Planter. Again failed to make contact on radio-telephone but someone waved from the bridge, which was friendly.

Wind strengthened to force 4 from South so we reefed the main. Still making 6 knots on a beam reach—excellent. Noon position—43°18' N, 33°22' W.

Slight freshening of the wind brought us down to No. 1 stays'l at 1500 and at the same time we replaced some more of the sliders on the mainsail bolt rope. Lots of sunshine but barometer falling slowly.

William Bligh now reenters our story, as the mentor of a young man who was to be one of the leading maritime explorers of his generation. Once the court-martial of the mutineers was over, Bligh was free to resume normal duties—and to finish the job he had started in the *Bounty*. In 1791 he sailed once again for Tahiti via the Cape of Good Hope and the south coast of Australia in the *Providence*. The ship's company this time included a midshipman named Matthew Flinders. Having collected the long-delayed supply of breadfruit seedlings, Bligh had to decide how best to deliver them to the West Indies. The most direct route—around Cape Horn—was ruled out on the grounds that the delicate tropical plants would not survive the cold. So Bligh sailed westward, but instead of taking the relatively safe and well-known passage around the north of New Guinea, he decided once again to thread his way through the Torres Strait.

This was audacious, given that he was now sailing not in a shallow-draft open boat, but in a far less maneuverable square-rigged ship. Apart from Cook, no one had attempted this feat since Torres himself in 1606, and there were as yet no detailed charts. The whole area is dotted with coral reefs and low islands, through which the tides run strongly, making it exceptionally hazardous for sailing ships. Flinders was later to write that perhaps "no space of 3½ degrees [of longitude] in length presents more dangers than the Torres Strait," but—taking all due care—Bligh sailed through it safely, brushing off fierce attacks by the natives of Papua New Guinea on the way.[1] Even today large parts of the Torres Strait—away from the main safe passages—remain uncharted.

The breadfruit were delivered to St. Vincent and Jamaica in good condition (though the experiment was not a success, as the African slaves, for whom they were intended, seem not to have liked them), and Flinders returned to England in 1793. He had assisted Bligh in preparing charts and making astronomical observations, but their relationship deteriorated during the latter part of the voyage. The reasons are not clear, though if, as some believe, Flinders—in common with more than a quarter of the crew—contracted a sexually transmitted disease in Tahiti, his commander would certainly not have been pleased.[2] In any case, Flinders was convinced that Bligh had a "prepossession" against him, and he later suspected him of taking credit for his labors.[3]

Flinders was born in Lincolnshire in 1774 and was inspired to become a sailor by reading Daniel Defoe's tale of Robinson Crusoe.[4] That he had already studied Euclid's geometry and two manuals of navigation by the time he joined the Royal Navy in 1789 at the age of fifteen is a measure of his determination to succeed. He served first as a midshipman in the famous seventy-four-gun *Bellerophon* under Captain Thomas Pasley, an influential figure who was to take a close interest in his career,[5] before being appointed to the *Providence*.[6]

After serving with Bligh, Flinders renewed contact with Pasley, and again joined the *Bellerophon*, this time as an aide-de-camp. In this capacity he saw action in the running battle with the French fleet known by the British as the Glorious First of June (1794). The *Bellerophon* was severely damaged and Pasley lost a leg, but Flinders himself survived the carnage unscathed: it was his only experience of naval warfare, and he emerged from the action with credit.[7] The appointment of John Hunter as governor of the recently established colony of New South Wales now gave him the opportunity to return to the land that would before long be known as Australia.[8]

In February 1795, Flinders—still only a midshipman—sailed with Hunter in the *Reliance*, reaching Port Jackson in September. He was, he later wrote, led "by his passion for exploring new countries, to embrace the opportunity of going out upon a station which, of all others, presented the most ample field for his favourite pursuit."[9] His devotion to the practice of celestial navigation was already well established and he kept a careful note of his personal sextant observations on the outward passage. Soon after arriving in Port Jackson he launched the first of a series of coastwise surveying expeditions with his new friend George

Bass, a fellow Lincolnshire man, who had sailed out with him in the *Reliance* as ship's surgeon.

Bass, aged thirty-two, was a commanding figure with a "penetrating countenance," and Flinders—no milksop himself—described him as "one whose ardour for discovery was not to be repressed by any obstacle or deterred by any danger." Together the two young men decided to complete as far as possible the exploration of the east coast of "New South Wales"—a term that was then applied indiscriminately to the whole eastern half of Australia. Initially there was no official support for this ambitious undertaking, but Bass had brought out from England an eight-foot dinghy, and it was in this ludicrously small craft—named *Tom Thumb*—that he and Flinders, accompanied by an unnamed "boy," set out from Port Jackson. It is hard to imagine how they all fit into such a small boat. Their first trip took them only as far as Botany Bay, but from there they travelled some distance up George's River, breaking new ground, and their "favourable report" to Governor Hunter led to the establishment of a new settlement called Banks' Town—in honor of Joseph Banks.[10]

A longer and even more adventurous boat trip in *Tom Thumb* followed in March 1796. This took them farther south, where they had an alarming encounter with a group of aborigines, who they feared might be cannibals. While their gunpowder—which had been soaked—was drying, Flinders entertained the natives by clipping their hair and beards as a way of avoiding trouble:

> *Some of the more timid were alarmed at a formidable instrument coming so near to their noses, and would scarcely be persuaded by their shaven friends, to allow the operation to be finished. But when their chins were held up a second time, their fear of the instrument,—the wild stare of their eyes,—and the smile which they forced, formed a*

> *compound upon the rough savage countenance, not unworthy of the pencil of a Hogarth.*[11]

Having narrowly escaped being lost in a gale, they returned safely to Port Jackson in April 1796. Official duties, including a voyage to the Cape of Good Hope by way of Cape Horn—a circumnavigation they treated almost as routine—kept both men busy for some time, but Bass was able to conduct further inland forays, and in December 1797 he obtained leave to make a longer seaborne expedition to the south, this time without Flinders. He set off in a twenty-eight-foot whaleboat, supplied and victualled by the governor, with a crew of six sailors. At this date it was unclear whether Van Diemen's Land (now Tasmania) was an island, and Bass headed south in the hope of settling the matter. By January 2, 1798, having already travelled well over four hundred nautical miles, he reached the southernmost point of Australia—later named Wilson's Promontory in honor of a friend of Flinders—but there were heavy seas and the boat was leaking badly. So instead of turning south toward the north shore of Van Diemen's Land, he clung to the mainland coast, taking sextant sights to fix his position. Eventually he reached Western Port, just failing to discover the wide bay of modern Melbourne, a few miles farther to the west.[12] The long southwesterly swell, and the strength of the tides, strongly suggested to Bass that the stretch of water between him and Van Diemen's Land was a strait that connected with the open sea to the west.[13] However, the poor condition of the boat and shortage of food obliged him to head for home, and he reached Port Jackson again on February 25 after enduring a series of heavy gales. Though the positions reported by Bass turned out to be inaccurate, Flinders excused him on the grounds that his sextant had probably been damaged.[14] He later wrote:

> *A voyage especially undertaken for discovery in an open boat, and in which six hundred miles of coast, mostly in a boisterous climate, was explored, has not, perhaps, its equal in the annals of maritime history.*[15]

The phrase "especially undertaken for discovery" was presumably a concession to the extraordinary boat journey of his former commander, Bligh.

Since Bass had not yet conclusively proved that the body of water now named after him was indeed a strait separating Van Diemen's Land from the mainland, in October 1798 Flinders and Bass—together once again—set off to settle the matter in a small twenty-five-ton sloop, the *Norfolk*, with a crew of eight volunteers. Unfortunately, it was impossible to obtain a chronometer, so Flinders had to rely entirely on lunars to determine their longitude.[16]

Making meticulous sextant observations to fix his position at every opportunity, and surveying every useful anchorage or harbor, Flinders led the expedition through the islands at the eastern end of the straits and along the northern shore of Van Diemen's Land, where they discovered Port Dalrymple[17] and explored the Tamar River (now the location of the town of Launceston).[18] When they reached the northwestern tip of the island, they encountered a long swell from the southwest, which they hailed "with joy and mutual congratulation, as announcing the completion of our long-wished-for discovery of a passage into the southern Indian Ocean."[19] Flinders sailed down the inhospitable west coast of Van Diemen's Land and spent some days exploring the Derwent River, at the head of the extensive natural harbor on the south coast discovered in 1792 by d'Entrecasteaux on his voyage in search of La Pérouse.[20] On the basis of Flinders's favorable report, a new colony was established there four years later, known today as Hobart. By January 12, 1799, the

Norfolk had returned to Port Jackson, and at Flinders's request the governor named the newly discovered strait after George Bass.[21] The new passage was of more than purely geographical interest: it significantly reduced the time taken by ships sailing out to Sydney* since they could now avoid the long detour around the south coast of Van Diemen's Land. The two friends were soon separated, and Bass—who had by then left the navy—disappeared in mysterious circumstances after undertaking a trading voyage from Port Jackson to South America in 1803. According to one rumor, he was captured by the Spanish colonial authorities and condemned to the mines on suspicion of smuggling. Whatever the truth, Bass was never heard of again.[22]

After conducting one more relatively brief survey voyage—this time of the coast north of Port Jackson—Flinders returned to England, arriving home in August 1800. By now he was a lieutenant, and his achievements had already brought him to the attention of the scientific world. The next and most important phase of his career began in January 1801, when, having won the powerful backing of Sir Joseph Banks, Flinders was given command of a small sloop—renamed the *Investigator*—to complete the survey of the entire coast of the continent for which he was to propose the name Australia.[23] The *Investigator* was "a north-country built ship, of three-hundred and thirty-four tons; and, in form, nearly resembled the description of vessel recommended by captain Cook as best calculated for voyages of discovery."[24] Flinders was only twenty-six years old and it was the job of his dreams, but there was a problem: he had fallen in love with a young woman who lived near his home in Lincolnshire, Ann Chappelle.

Flinders was determined to make a name for himself, but he

---

* The return trip would usually be via Cape Horn, taking advantage of the westerly winds that prevail in the mid-latitudes of the South Pacific.

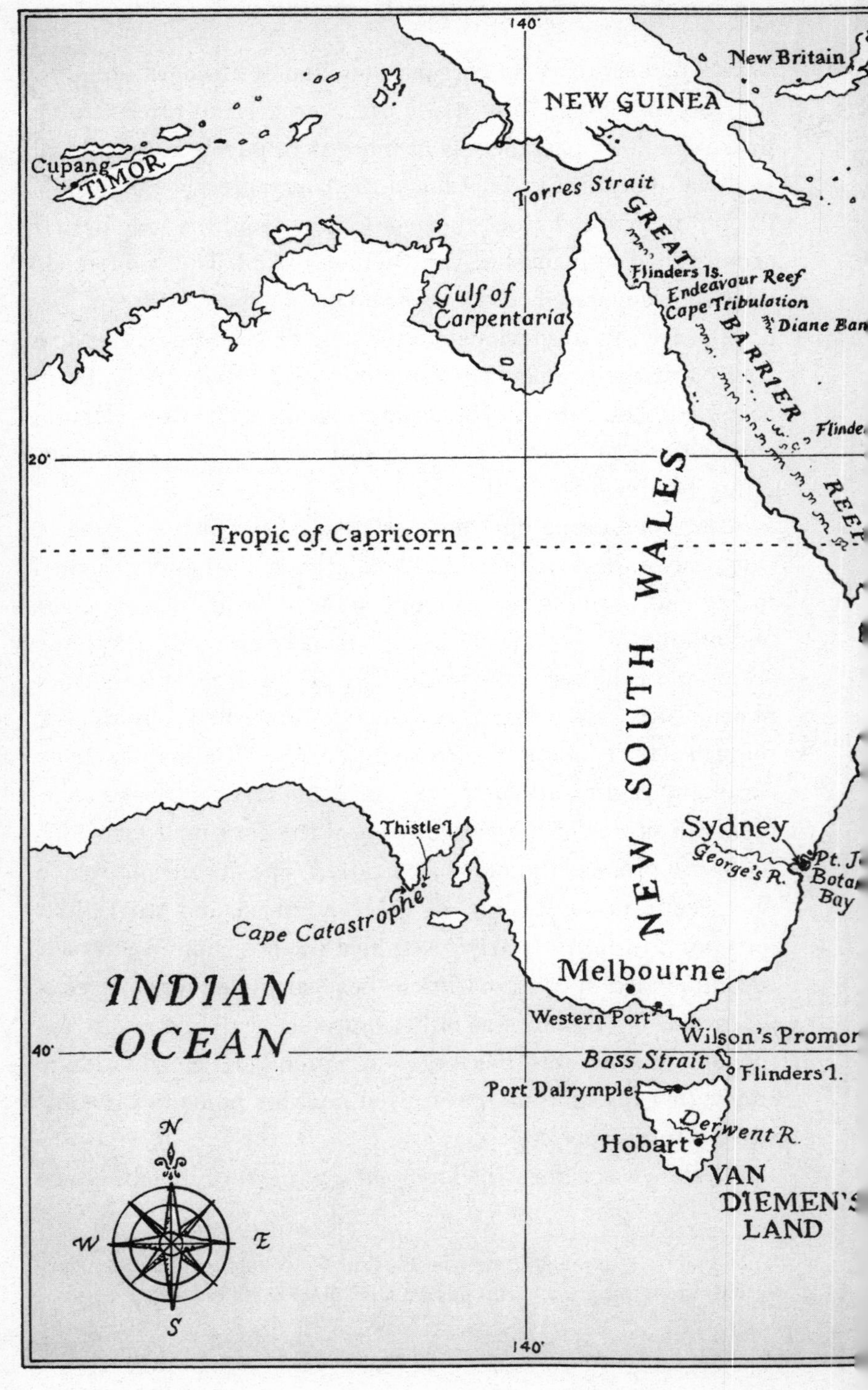
140°
New Britain
NEW GUINEA
Cupang
TIMOR
Torres Strait
GREAT
Flinders Is.
Endeavour Reef
Cape Tribulation
Gulf of Carpentaria
BARRIER
REEF
20°
Tropic of Capricorn
NEW SOUTH WALES
Thistle I.
Cape Catastrophe
Sydney
George's R.
Bay
INDIAN OCEAN
Melbourne
Western Port
40°
Bass Strait
Flinders I.
Port Dalrymple
Derwent R.
Hobart
LAND
N
W
E
S
140°

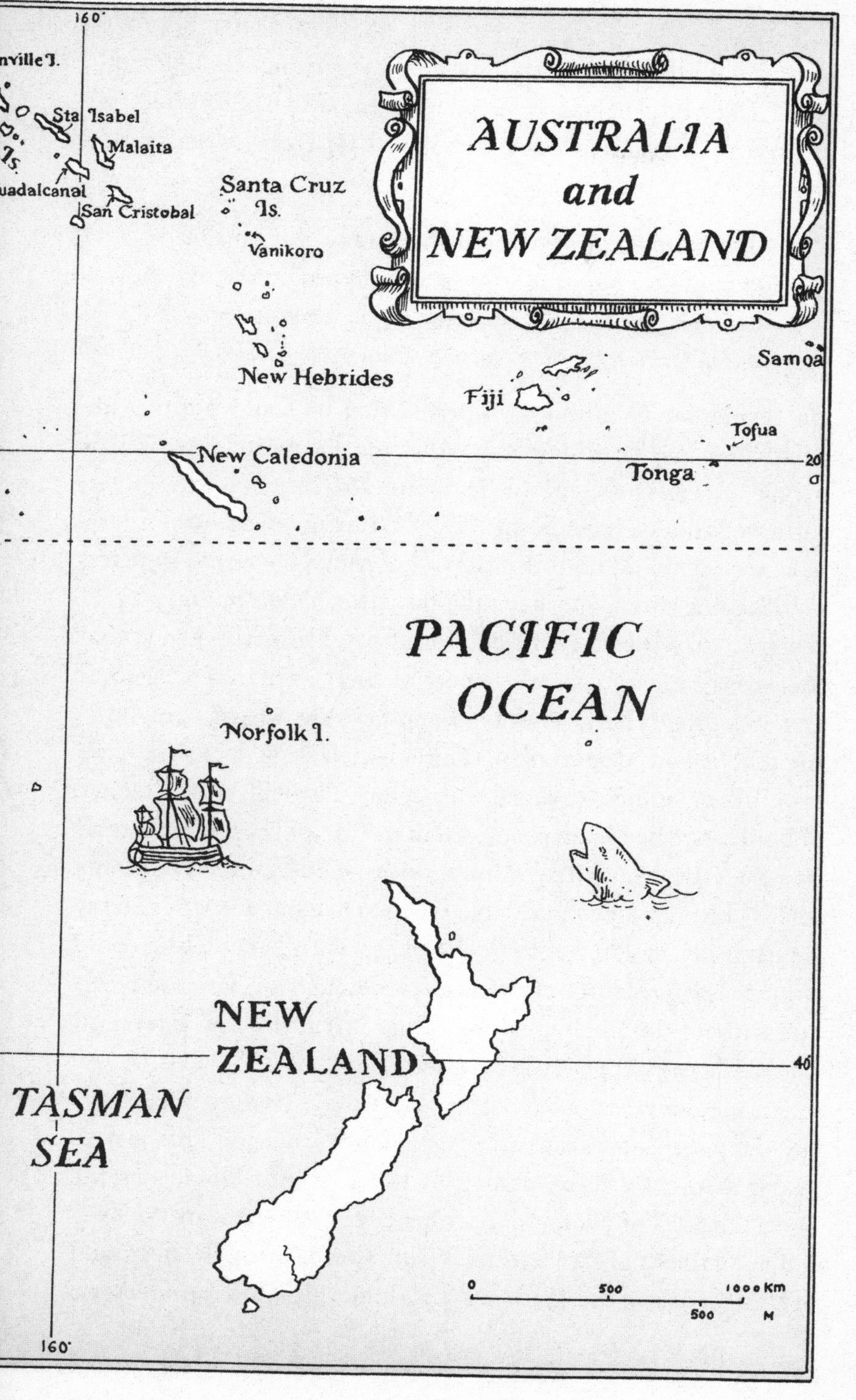
AUSTRALIA
and
NEW ZEALAND
160°
nville I.
Sta Isabel
Malaita
Is.
uadalcanal
San Cristobal
Santa Cruz
Is.
Vanikoro
New Hebrides
Fiji
Samoa
Tofua
Tonga
New Caledonia
20
PACIFIC
OCEAN
Norfolk I.
NEW
ZEALAND
40
TASMAN
SEA
0
500
1000 Km
500
M
160°

also wanted to marry Ann: how could the two goals be reconciled? His troubled state of mind is revealed in his letters. After an argument with his father over money, he wrote to Ann on November 29, 1800:

> *I have thrown away what might have been the means of making thee happy . . . my soul is turned inside out to thee:—fear not to equal my openness. . . . Fortune may favour me, or may turn her back upon me—Heaven knows.*[25]

On December 18, Flinders, worried that he could not provide adequately for her, told Ann that they must remain "two distinct people. . . . Let us then, my dear Annette, return to the 'sweet, calm delights of friendship.' . . ."[26] After meeting her on January 16, 1801, and doubtful now of her love, he wrote in tears, bidding her adieu, "perhaps the last time."[27] On January 27, in response to a direct question from Ann, Flinders warned that there was only a distant prospect of their ever living together and encouraged her to seek her happiness elsewhere, complaining that he had "undertaken a Herculean task."[28] By April, however, his mood had changed completely. Through the influence of Banks[29] he had been promoted to the rank of commander and now saw "the probability of living with a moderate share of comfort."[30] Flinders therefore asked Ann to become his wife, rashly inviting her to sail out to Port Jackson with him—if she could support the hardships of the passage. Though he expressed concerns about the likely reaction of his "great [that is, powerful] friends," they were swiftly married.

An incident that occurred in May, when Flinders was bringing the *Investigator* around to Portsmouth from the Thames estuary, discreetly accompanied by his new wife, illustrates the shortcomings of contemporary hydrography—even in the busy waters of the English Channel. While approaching the headland of Dungeness on the south coast of Kent, the ship grounded on a sandbank known locally as "the Roar."[31] This hazard was marked on most contemporary charts, but not, as bad luck would have it, on that with which Flinders had been supplied.[32] The conditions were calm, there was a rising tide, and the *Investigator* suffered no damage before she floated off, but Flinders was deeply embarrassed.

By now the Admiralty had discovered that he had brought his wife aboard, and their lordships took the view that the grounding was due not to a faulty chart, but to the distracting effects of Mrs. Flinders. Ann might perhaps have been able to accompany Flinders on the long outward voyage to Port Jackson—such arrangements were not unheard-of aboard men-of-war when permission had been obtained. However, it would have been unprecedented for the commander of a vessel on a voyage of discovery to take his wife with him, and in any case Flinders had not even sought official approval. In the face of formal complaints from the Admiralty, and the stern disapproval of Banks, Flinders had no choice but to leave behind the woman he had just married. Writing from Portsmouth on June 30, he told Ann: "when thy remembrance assails me too powerfully and my eyes swell at the recollection, I shall retire to *the little bed* [Flinders's emphasis], and vent my sorrow to myself alone."[33] They were not to meet again for nine years.

Like Cook's *Endeavour*, the *Investigator* was a flat-bottomed vessel capable of taking the ground safely, though as soon as she was at sea the extent of the leaks revealed that her hull was in poor condition. The ship's company numbered eighty-eight, and the crew had all volunteered—evidence perhaps of the glamour that surrounded voyages of discovery, and of the high reputation that Flinders had already won. Flinders's younger brother Samuel sailed as second lieutenant, and one of the six midshipmen was John Franklin, who was to achieve posthumous fame in his disastrously unsuccessful bid to find the Northwest Passage through the Arctic in the 1840s. The "supernumeraries" included the landscape artist William Westall, only nineteen at the time, as well as Ferdinand Bauer, who was to make some of the finest botanical and zoological drawings of all time while on the voyage, and Robert Brown, a gifted young naturalist selected by Banks. The East India Company, which had a monopoly on British trade in the Far East and stood to benefit from the navigational data deriving from the voyage, supplied £600 toward the costs of the officers' food and drink. In addition to the crew there were plenty of livestock: sheep, goats, pigs, and chickens—not to mention two cats, including the captain's favorite, Trim, who had been born aboard the *Reliance* and had accompanied him on his earlier surveying voyages.

The *Investigator* set sail from Spithead (the stretch of water between Portsmouth and the Isle of Wight) on July 18, 1801, and after calling at Madeira and the Cape of Good Hope, made her landfall on Cape Leeuwin at the southwestern corner of New Holland on December 6. The "Herculean task" that Flinders had foreseen was now to begin. His instructions from the Admiralty were demanding: "You are to be very diligent in your examination of the [Australian] coast, and to take particular care to insert in your journal every circumstance that may be useful to a full and complete knowledge thereof, fixing in all cases, when in your power, the true positions both in latitude and longitude of remarkable headlands, bays, and harbours, by astronomical observations."[34] Many parts of this coast were still uncharted, and there was considerable uncertainty about the overall shape of the continent—in particular whether it might be divided into two halves, or perhaps even embrace a large inland sea.* Flinders was therefore required to explore any riv-

* Ignorance of the geography of the Southern Hemisphere was exploited for literary purposes by Jonathan Swift, whose Gulliver was shipwrecked on the shores

ers and "examine the country as far inland as shall be thought prudent . . . discovering any thing useful to the commerce and manufactories of the United Kingdom." Nor was he to forget the needs of science and art: "During the course of your survey, you are to [move] the *Investigator* onward from one harbour to another as they may be discovered, in order that the naturalists may have time to range about and collect the produce of the earth, and the painters allowed time to finish as many of their works as they possibly can on the spot where they may have been begun."[35]

Flinders needed no official encouragement: he had a scientific bent and was determined to raise the standards of hydrography to the highest possible level. Having already shown himself a skillful and daring navigator, over the next two years he painstakingly charted roughly half the coast of Australia. He corrected existing surveys—including, apologetically, some details of those made by the great Cook—and extended them into little-known regions, though he was disappointed not to discover any major inlet that would give access to the interior.[36] As far as possible he avoided the "running survey" method. Instead he anchored frequently, and when he went ashore used the theodolite to connect fixed positions by triangulation. He relied heavily on chronometers for his survey work but checked them as often as possible by lunars and sometimes, when ashore, by the moons of Jupiter. On one occasion he observed an eclipse, which proved a useful index of the accuracy of his methods after he had returned home.

Flinders also made an extended series of observations of the strangely erratic behavior of magnetic compasses aboard ship.

---

of Lilliput after being driven by a "violent storm to the north-west of Van Diemen's Land" (chapter 1 of *Gulliver's Travels*, which was first published in 1726). This would place Lilliput somewhere within the continent of Australia.

Their tendency to give different bearings of the same object depending on which way the ship's head was pointing had long been a mystery and was a matter of great navigational significance. Flinders realized that this "compass deviation" was due to the influence of magnetic metals aboard the ship, and he devised a means of correcting it that was tested successfully after his eventual return to England. Though it was many years before his proposal was adopted, compensatory "Flinders bars" eventually became a standard feature of steering-compass installations on board every modern ship. Flinders, who kept careful records of the barometric pressure, was also among the first to investigate carefully the relationship between changes in this variable and the behavior of the wind, a study that was to be taken further by another distinguished hydrographer—Robert FitzRoy.

Like so many other voyages of exploration, this one was marked by human tragedy. In February 1802 the master of the *Investigator*, John Thistle, along with seven other men, was drowned on an expedition in one of the ship's boats in rough waters off the south coast of Australia. Thistle, who had earlier sailed with Flinders and Bass, was both an old friend and a much-valued colleague. The site of the accident was appropriately called Cape Catastrophe and the grief-stricken Flinders named an island after Thistle.

While the fruitless search for the lost men was under way, Flinders learned that Thistle before leaving home had visited a fortune-teller who had told him that he was going on a long voyage and that his ship would be joined by another vessel after reaching her destination; he, however, would be lost before that happened.[37] The superstitious crew of the *Investigator* were very struck by this prophecy, but Flinders coolly observed that other commanders should discourage their crews from consulting fortune-tellers.

Oddly enough, they did have an unnerving chance encounter with another vessel some weeks later. This was *Le Géographe*, commanded by the French explorer Nicolas Baudin, who was proceeding along the south coast in the opposite direction. Uncertain of the Frenchman's intentions, the little *Investigator* cleared for action and veered around as she passed the larger ship in order to keep her broadside to her. Flinders then hove to and went aboard *Le Géographe*. It seems to have been a slightly tense encounter. Baudin's lack of interest in the identity of his interlocutor, or even in his reasons for being in this remote part of the world, plainly nettled Flinders, who could be touchy:

> *Captain Baudin was communicative of his discoveries about Van Diemen's Land; as also of his criticisms upon an English chart of Bass Strait published in 1800. He found great fault with the north side, but commended the form given to the south side and to the islands near it* [surveyed by Flinders]. *On my pointing out a note upon the chart, explaining that the north side of the strait was only seen in an open boat by Mr. Bass, who had no good means of fixing either latitude or longitude, he appeared surprised, not having before paid attention to it.*[38]

The two ships kept company overnight, and the following day Flinders told Baudin, in guarded terms, what the *Investigator* had been doing, as well as offering him advice about places where he could obtain freshwater and food. Only when the time came to part did Baudin think to ask Flinders for his name:

> *and finding it to be the same as the author of the chart he had been criticising,* [he] *expressed not a little surprise, but had the politeness to congratulate himself on meeting me.*[39]

In May 1802 Flinders called at Port Jackson and seized the opportunity to write home to Ann:

> *A moment snatched from the confusion of performing half a dozen occupations, and of making up eighteen months accounts in every one of them, is a poor tribute to offer to a beloved friend like thee. That I am safe and well, and have done everything thus far that I could have expected to do, is to tell thee something. How highly should I value such short information reciprocated from thee! but alas, my dearest love, I am all in the dark concerning thee, I know not what to fear or what to hope. Pray write and releive* [sic] *my anxiety.*

Flinders reported that Ann would have been comfortable at Port Jackson, and accused "fortune of great unkindness" for denying him the joys of her company. He touched also on financial matters, explaining that he had sent money home to his agent to the value of almost £400:

> *Thou wilt judge from the above, that notwithstanding the arduous task of being astronomer, surveyor, commander, and inspector of every officer and mans conduct and accounts, that my pecuniary concerns have not been neglected. No, my beloved, thou art concerned in these, and I shall not cease to do every thing for thee, until life, or the requisite power, ceases. I still think that the voyage will be as beneficial to us, as I ever supposed; which was, that I should be fifteen hundred pounds richer at the conclusion than at the commencement of it; this, however, need not be said to every body.*[40]

At Port Jackson he again encountered *Le Géographe*, whose crew had been reduced to a desperate state by scurvy—only twelve out of 170 men were still fit for duty.[41] The governor, with the active assistance of Flinders, did all in his power to help the French sailors, and a dinner was held on board the *Investigator* to celebrate the news that Britain and France were no longer at war, and to which Baudin and his officers were invited. By now two of the chronometers available to Flinders had stopped, and he was henceforth reliant on the two (by Earnshaw) that were still in good order—a scarcely adequate provision for the challenging survey work that lay ahead along the Great Barrier Reef and the little-known north coast.[42] As Flinders proceeded up the coast of what is now Queensland, La Pérouse, who had by then been missing for almost ten years, was much in his thoughts:

> *At every port or bay we entered . . . my first object on landing was to examine the refuse thrown up by the sea. The French navigator, La Pérouse, whose unfortunate situation, if in existence, was always present to my mind, had been wrecked, as it was thought, somewhere in the neighbourhood of New Caledonia; and, if so, the remnants of his ships were likely to be brought upon this coast by the trade winds, and might indicate the situation of the reef or island which had proved fatal to him. With such an indication, I was led to believe in the possibility of finding the place; and though the hope of restoring La Pérouse or any of his companions to their country and friends could not, after so many years, be rationally entertained, yet to gain some knowledge of their fate would do away the pain of suspense: and it might not be too late to retrieve some documents of their discoveries.*[43]

Flinders found no clues and carried on to make detailed observations in the Torres Strait before turning south to explore the Gulf of Carpentaria. This was arduous, time-consuming, and not very rewarding: there were no proper harbors, no rivers of any significance—in fact, there was nothing much of use to anyone, apart from the occasional turtle to brighten up the dinner menu. The climate was oppressively hot and humid (one man died of sunstroke), there were occasional clashes with the natives, and they were plagued by mosquitoes and flies. Reading between the official lines it is also clear that Flinders's younger brother, Samuel, was a trial: on several occasions he committed the grave sin of allowing the two chronometers to run down, thereby necessitating lengthy delays while they were freshly rated. Charting the whole gulf coast with such care must have been gruelling work, but little flashes of humor nevertheless brighten up Flinders's account. He came across some small red crabs, with one claw nearly as big as their bodies (presumably some kind of fiddler crab), and was amused "to see a file of these pugnacious little animals raise their claws at our approach, and open their pincers ready for an attack; and afterwards, finding there was no molestation, shoulder their arms and march on."[44]

In November 1802, while on the gulf coast, Flinders took the precaution of careening the leaky *Investigator* and made the shocking discovery that her hull was in a state of almost terminal decay. A written report from the master and the carpenter left him in no doubt that the voyage would have to be curtailed:

> *I cannot express the surprise and sorrow which this statement gave me. . . . My leading object had hitherto been to make so accurate an investigation of the shores of Terra Australis that no future voyage to this country should be necessary; and with this always in view, I had ever endeavoured to follow the land so closely that the washing of the surf upon it should be visible, and no opening, nor anything of interest, escape notice. Such a degree of proximity is what navigators have usually thought neither*

> *necessary nor safe to pursue, nor was it always persevered in by us; sometimes because the direction of the wind or shallowness of the water made it impracticable, and at other times because the loss of the ship would have been the probable consequence of approaching so near to a lee shore. But when circumstances were favourable, such was the plan I pursued, and, with the blessing of GOD, nothing of importance should have been left for future discoverers upon any part of these extensive coasts.*[45]

Now, however, Flinders was in no doubt that the *Investigator* could not survive heavy weather at sea, and that there would be no hope of repairing her if she sustained serious damage on any of the "numerous shoals or rocks upon the coast." Even if constant fine weather could be relied on and all accidents avoided, the ship could not be expected to stay afloat for more than six months. Flinders concluded bitterly that "with such a ship" he had no chance of accomplishing the task he had been set: he had no choice but to return to Port Jackson in the forlorn hope of finding there a new vessel in which to complete his mission.

Flinders sailed via the west coast, choosing not to risk the much shorter but more hazardous route back through the Torres Strait in the teeth of the strong northeasterly trade winds. Shortage of fresh provisions obliged him to stop at Cupang, which—like Batavia—was a most unhealthy place. By the time he reached Port Jackson six months later, in June 1803, having thereby completed the first circumnavigation of Australia, the ship and many of her crew were in a desperate state. Several men had already died of dysentery picked up in Timor, many others were seriously ill, and Flinders himself was so disabled by sores caused by scurvy that he could no longer go aloft—as his surveying work frequently required. An examination of the *Investigator* revealed that there were thirteen timbers in her starboard bow "through any of which a cane might have been thrust." Flinders reflected that during their recent passage along the south coast of Australia, the strong southerly breezes had providentially heeled the ship to port. Had the wind set in from the north, "the little exertion we were then capable of making at the pumps could hardly have kept the ship up; and a hard gale from any quarter must have sent us to the bottom."

Writing to Ann, Flinders thanked her for the letters that had greeted him on his arrival at Port Jackson, and gave her news of the further deaths among the crew, his own ill health, and all the problems he now faced. He then poured out his heart to her:

> *Thou hast shewn me how very ill I have requited thy tender love in several instances. I cannot excuse myself, but plead for respite until my return when in thy dear arms I will beg for pardon, and if thou canst forgive me all, will have it sealed—oh with ten-thousand kisses. . . .*

Implicitly acknowledging that his ambition had kept them apart, Flinders apologetically assured Ann that

> *so soon as I can insure for us a moderate portion of the comforts of life, thou wilt see whether love or ambition have the greatest power over me. Before thou wast mine, I was engaged in this voyage;—without it we could not live. Thou knowest not the struggle in my bosom, before I consented to the necessity. There was no prospect of a permanent subsistence but in pursuing what I had undertaken, and I doubt not but that it will answer its end.*[46]

Chapter 13

Flinders—Shipwreck and Captivity

Day 14: A grey dawn but still a fair wind and making 5 knots. Colin announced that we were halfway. The bread from Halifax has now turned almost entirely green so breakfast was scrambled eggs mixed with the remaining edible crumbs—surprisingly good. Seas still pretty lumpy so we stayed under No. 1 stays'l. Barometer still falling.

Colin complimented me on my sextant work today and let me do the mer alt and noon fix on my own: 44°23' N, 30°36' W.

Depression came through in afternoon. Low cloud, wind going round into SW and then rain and strong squalls as the warm front moved through. Force 6 to 7 but it quickly kicked up an uncomfortable sea. Wind later veered to W and started to ease though skies still thick with cloud. Finally, during the night, the wind veered to N.

During night Colin and I were up and down making several sail changes. Little sleep.

Having recovered his strength, and consulted the governor—Philip Gidley King (1758–1808)—Flinders set sail again in August 1803 aboard the small armed vessel *Porpoise*, in the hope of obtaining a new survey ship when he reached England, no suitable vessel being available in Port Jackson. The *Porpoise* was accompanied by two larger ships, both merchantmen—the *Cato* and the *Bridgewater*. On the way Flinders planned to extend his earlier examination of the Torres Strait and also to demonstrate that this route was reasonably safe by leading the two other ships through it. But luck was not on Flinders's side. Long before they reached the Torres Strait, the *Porpoise* and the *Cato* were wrecked on an isolated and uncharted coral reef lying several hundred miles off the coast of Queensland. As Flinders was technically a passenger, he was not called when, during the night, breakers were seen ahead. By the time he reached the deck it was too late to save the ship:

> *On going up, I found the sails shaking in the wind, and the ship in the act of paying off; at the same time there were very high breakers at not a quarter of a cable's length* [150 feet] *to leeward. In about a minute, the ship was carried amongst the breakers; and striking upon a coral reef, took a fearful heel over on her larboard beam ends. . . .*
>
> *Our fore mast was carried away at the second or third shock; and the bottom was presently reported to be stove in, and the hold full of water. When the surfs permitted us to look to windward, the Bridgewater and the Cato were perceived at not more than cable's length distance* [200 yards]; *and approaching each other so closely that their running abord* [sic] *seemed to us inevitable. This was an aweful moment: the utmost silence prevailed; and when the bows of the two ships went to meet, even respiration seemed suspended. The ships advanced, and we expected to hear the dreadful crash; but presently they opened off from each other, having passed side by side without touching. . . . Our own safety seemed to have no other dependence than upon the two ships, and the exultation we felt at seeing this most imminent danger passed, was great, but of short duration; the Cato struck upon the reef about two cables length from the Porpoise, we saw her fall over on her broadside, and the*

> *masts almost instantly disappeared; but the darkness of the night did not admit of distinguishing, at that distance, what further might have happened.*[1]

The *Bridgewater* narrowly escaped Wreck Reef, but her master, Captain Palmer, to the disgust of his shipmates, made no serious attempt to help the survivors of the two wrecks, claiming on his arrival in India that they must all have been lost. The third mate of the *Bridgewater* challenged his account, and he, as well as several other officers, then quit the ship. It was lucky for them that they did: Palmer and the *Bridgewater* never reached home. "How dreadful must have been his reflexions at the time his ship was going down!" Flinders later commented dryly.[2]

Luckily the survivors were able to salvage a good deal from the *Porpoise*—including a sextant, three timekeepers, and Flinders's log and bearing books, as well as most of his charts and astronomical observations—and they managed to reach a dry sandbank within the reef where they pitched camp. Two of the timekeepers had stopped, but one was still going well.

When searching for firewood on the first night after they landed, they found a spar and a piece of timber, worm-eaten and almost rotten. The timber was seen by the master of the *Porpoise*, who judged it to be part of the stern-post of a ship of about four hundred tons; and Flinders thought it might, "not improbably, have belonged to *La Boussole* or *L'Astrolabe*":

> *Monsieur de la Pérouse, on quitting Botany Bay, intended to visit the south-west coast of New Caledonia; he might have encountered in the night, as we did, some one of the several reefs which lie scattered in this sea. Less fortunate than we were, he probably had no friendly sand bank near him . . . or perhaps the two vessels both took the unlucky direction of the Cato after striking, and the seas which*

broke into them carried away all his boats and provisions; nor would La Pérouse, his vessels, or crews be able, in such a case, to resist the impetuosity of the waves more than twenty-four hours. If such were the end of the regretted French navigator, as there is now but too much reason to fear, it is the counterpart of what would have befallen all on board the Porpoise and Cato, had the former ship, like the Cato, fallen over towards the sea instead of heeling to the reef.[3]

With characteristic determination, Flinders took charge of the ninety-odd survivors. They had enough food and water to last three months, so Flinders set off on August 26, 1803, in one of the surviving boats (a six-oared cutter with a crew of fourteen in all, given the name *Hope*) to get help from Port Jackson, 739 nautical miles away. This was a long and perilous journey to make in an open boat, even if it did not match the scale of the voyage Bligh had been forced to undertake. As they pushed off, Flinders recalled, a seaman ran to the makeshift flagstaff on the sandy island, hauled down the ensign, which had earlier been hoisted upside down as a distress signal, and rehoisted it right side up: "This symbolical expression of contempt for the *Bridgewater* and of confidence in the success of our voyage, I did not see without lively emotion."[4]

The *Hope* was overladen and soon encountered steep seas that caused her to labor so much that Flinders was forced to lighten her by throwing overboard some of the freshwater, food, and cooking apparatus. On the fourth day, however, they made their landfall on the coast and on September 8 they reached Port Jackson.

The reader has perhaps never gone 250 leagues at sea in an open boat, or along a strange coast inhabited by savages;

> *but if he recollect the 80 officers and men upon Wreck-Reef Bank, and how important was our arrival to their safety, and to the saving of the charts, journals, and papers of the Investigator's voyage, he may have some idea of the pleasure we felt, but particularly myself, at entering our destined port.*[5]

On his unexpected return, Flinders went at once to the governor's house to organize a rescue mission. Bursting in on King and his family when they were dining, the bearded and sunburned Flinders made a never-to-be-forgotten impression on King's young son, Phillip.

Only six weeks after leaving the reef, Flinders returned to Wreck Reef aboard a schooner, the *Cumberland*, accompanied by a much larger merchant vessel, the *Rolla*, which happened to have called at Port Jackson on her way to Canton, and another small craft, the *Francis*:

> *On landing, I was greeted with three hearty cheers, and the utmost joy by my officers and people; and the pleasure of rejoining my companions so amply provided with the means of relieving their distress made this one of the happiest moments of my life.*[6]

The survivors had by then almost given up hope of seeing Flinders again and had already built, out of timbers gathered from the wrecks, a small, decked sailing boat in which they planned to save themselves. Samuel Flinders had devoted much time to fixing the position of the reef by means of lunars. All were now rescued, most taking passage in the *Rolla*, while some chose to return to Port Jackson either in the *Francis* or in the newly built boat.

Flinders himself set sail from Wreck Reef for England in the

Cumberland, which was more of a yacht than a ship. She was not a good choice, and Flinders candidly admitted that he was influenced by the wish to be the first to undertake such a long voyage in so small a vessel.[7] His crew consisted of ten volunteers from the crew of the *Investigator*, and he took with him his charts and books, his personal "specimens of mineralogy and conchology," as well as the instruments originally supplied by the Navy Board and "the sole time keeper which had not stopped." Still eager to gather hydrographic data, he sailed again through the Torres Strait, but he soon discovered that the *Cumberland* leaked badly, and that her two pumps were in a poor state of repair: in fact, one pump gave up almost completely after they had entered the Indian Ocean. In these circumstances Flinders recognized that it was far too risky to attempt rounding the Cape of Good Hope—where the strong westerly winds encountering the south-going Agulhas Current throw up steep and dangerous seas—so he decided to seek help from the French authorities on the Isle de France (Mauritius).[8] This proved to be a disastrous decision: unknown to him, Britain and France were again at war.

Flinders was carrying a French passport designed to give the *Investigator* immunity from any act of war because she was engaged in peaceful, scientific work. However, he was not sailing in the *Investigator* any longer, and the passport could therefore be deemed invalid. All might have been well, but the exhausted Flinders was disgusted by the suspicious and unfriendly behavior of the French authorities: the contrast with the generous manner in which the British had treated Baudin and his men when they arrived in Port Jackson in 1802 made him all the more furious.

Nor did it help that the recently arrived governor of the Isle de France, General Charles de Caen, had been frustrated in his perfectly legal attempt to reestablish a French military presence in India during the recent peace. He was a distinguished soldier, whose career had prospered under Napoleon, but now he was marooned on an island in the middle of the Indian Ocean, far from the action in Europe, with fading prospects and very little to do. De Caen found it hard to believe the astonishing story Flinders told him and seems to have suspected him of espionage. He was also annoyed that Flinders failed to remove his hat in his presence. Unfortunately the captain of *Le Géographe*—who would no doubt have vouched for Flinders—had sailed just a few days earlier.[9] Nevertheless, in a conciliatory spirit, de Caen later sent word for Flinders to join him and his wife for dinner. Every tragic hero has a flaw, and Flinders now revealed his by indignantly turning down the invitation. This was not the best time to indulge his prickly sense of honor, and he was to pay a very high price for his disastrous error of judgment. He later learned that if he had only swallowed his pride he might soon have been released; instead, the affronted de Caen decided to teach the uppity British naval officer a lesson.[10]

The heartbreaking result of this spat was that Flinders spent the next six and a half years on the Isle de France as a virtual prisoner, desperately trying by every means at his disposal to persuade the French authorities to let him return home—though his sarcastic letters to the recalcitrant de Caen cannot have helped his cause. Among those in France who petitioned on his behalf was the elderly Bougainville, who, having narrowly avoided the guillotine during the Revolutionary Terror, had found favor with Napoleon and was now in an influential position.[11] After an initial period of close confinement, Flinders was eventually allowed to move to a plantation in the highlands, where he lived with a French family who made him very much at home. He took advantage of his enforced leisure to draw the charts of his voyage, to learn French and even Malay—still hopeful that one day he might be able to finish the great task he had begun. He also had the opportunity to visit the estate, near the center of the island, where La Pérouse had lived during his stay on the island in the 1770s:

> *I surveyed [the estate] with a mixture of pleasure and melancholy. How happy he had once been in this little spot with his family, and what a miserable fate terminated his existence. This was the spot where the man lamented by the good and well-informed of all nations, the man whom science illumined, and humanity, joined to an honest ambition, conducted to the haunts of remote savages; in this spot he once dwelt unknown to the great world, but happy. When he became great and celebrated, he had ceased to exist. But his labours have not been taken in vain; by his foresight a part at least of the produce of them has been saved to the public, and his example will serve to animate the sincere lovers of science. . . .*[12]

Flinders arranged for a simple memorial to be placed on the spot where the house had stood, inscribed simply "La Pérouse." In 1897 it was replaced by a large conical rock on which Flinders's own tribute was quoted.

In March 1806 the seventy-seven-year-old Bougainville personally intervened on Flinders's behalf with Napoleon, and shortly afterward the French government gave the order for his release—"out of a pure sentiment of generosity."[13] But circumstances, including, ironically, a British naval blockade, continued to conspire against Flinders. His correspondence with Ann (of which sadly only his side survives) was inevitably intermittent during the period of his captivity, and this contributed to his sense of isolation. A letter sent to her in December 1806 reveals that he had been deeply depressed:

> *It is now my dearest friend fourteen months since the date of thy last letter received here. I much fear that some subsequent ones have been lost or detained. . . . The arrival of thy letters forms the greatest epochs in my present monotonous life, and I sigh for them as for the most desired of blessings; next to the liberation which should permit me to fly to thy arms they afford me the greatest happiness I can receive. . . . Cease not then, my best beloved, to write often if thou wouldst preserve me from distraction. A six months longer silence, without such an increase in my prospects as to give me the strongest assurance of obtaining liberty may, alas,—be productive of disastrous consequences. Such an accession of despair as I experienced in September, would be more than my mind could support.*[14]

Flinders told Ann that he had even contemplated the drastic step of breaking his parole and making his escape but had rejected the plan because he wanted his honor to remain "unstained." He was not in fact allowed to leave the Isle de France until June 1810:

> *after a captivity of six years, five months and twenty-seven days, I at length had the inexpressible pleasure of being out of the reach of general De Caen.*[15]

The scientific rigor of Flinders's approach to chart-making was unprecedented, as the published account of his voyage reveals:

> *Longitude is one of the most essential, but at the same time least certain data in hydrography; the man of science therefore requires something more than the general result of observations before giving his unqualified assent to their accuracy, and the progress of knowledge has of late been such, that a commander now wishes to know the founda-*

> *tion upon which he is to rest his confidence and the safety of his ship.*[16]

Flinders accordingly listed the precise results of the observations by which the longitudes of the most important points on each coast had been fixed, as well as the means used to obtain them. Lunar distances had been particularly important in regulating the timekeepers on which Flinders generally relied for his longitudes:

> *The instruments used in taking the distances, were a nine-inch sextant by Ramsden, and three sextants of eight inches radius by Troughton, the latter being made in 1801, expressly for the voyage . . . and each longitude is the result of a set of observations, most generally consisting of six independent sights. They were taken either by Lieutenant Flinders [Matthew's brother] or by myself. . . .*[17]

He explains the many corrections applied to these sights and the distances derived from them, including allowances for the "spheroidal figure of the earth" in accordance with the "very latest theory." But, he adds, these longitudes still required an even more important correction:

> *The theories of the solar and lunar motions not having reached such a degree of perfection as to accord perfectly with actual observation at Greenwich, the distances calculated from these theories and given in the almanack become subject to some error, and consequently so do the longitudes deduced from them.*[18]

By the time Flinders finally reached London in 1810, he had begun writing up his account of the *Investigator*'s voyage, and "the charts in particular were nearly ready for the engraver." However, "it was desirable that the astronomical observations, upon which so much depended, should undergo a re-calculation and the lunar distances have the advantage of being compared with the observations made at the same time at Greenwich."[19] Flinders therefore obtained permission for John Crosley, the official astronomer originally assigned to the *Investigator*, to compare all the predicted positions of the sun and moon on which the longitude calculations were based with those actually recorded on the relevant dates at the Royal Observatory[20]—with the not always enthusiastic assistance of Samuel Flinders. It turned out that the Greenwich observations of the two bodies differed so much from "the calculated places . . . given in the Nautical Almanacks of 1801, 2 and 3" as to require "considerable alterations in the longitudes of places settled during the voyage; and a reconstruction of all the charts. . . ."[21] Flinders wanly observed that, following the introduction of new, improved tables during his absence, "the necessity of correcting for errors in the distances at Greenwich will have ceased, or be at least greatly diminished." He also made some interesting comments about the degree of accuracy achievable using lunars:

> *some sea officers who boast of their having never been out more than 5', or at most 10', may deduce from the column of corrections in the different tables, that their lunar observations could not be entitled to so much confidence as they wish to suppose; since, allowing every degree of perfection to themselves and their instruments, they would probably be 12', and might be more than 30' wrong.*[22]

On the other hand, observers using the new tables might, Flinders predicted, achieve an error of 30' "on either side" from a single set of observations, but probably not more than 12', while sixty sets would "probably give the longitude exact to 1' or 2'." This degree of accuracy, he rightly noted, was far beyond the hopes of the first proposers of the lunar method. Everything, however, in the end depended on the accuracy of the tables:

> *In appreciating the degrees of accuracy to which a small or larger number of lunar distances may be expected to give the longitude, I suppose the observer to be moderately well practised, his sextant or circle,* and time keeper to be good, and his calculations to be carefully made; and it is also supposed, that the distances in the nautical almanack are perfectly correct. As, however, there may still be some errors, notwithstanding the science and labour employed to obviate them, it cannot be too much recommended to sea officers to preserve all the data of their observations; more especially such as may be used in fixing the longitudes of places but little, or imperfectly known. The observations may then be recalculated, if requisite . . . and the observer may have the satisfaction of forwarding the progress of geography and navigation. . . .*[23]

Although Flinders was briefly fêted on his return, his long-overdue promotion to the rank of post-captain was not backdated (for that familiar bureaucratic reason: the fear of setting a precedent), and he found himself short of money. A daughter was born in 1811, but he was already a sick man. A urinary-tract complaint, which had first manifested itself aboard the *Investigator* and may have resulted from gonorrhea contracted twenty years earlier in Tahiti,[24] worsened while Flinders was preparing his journal for publication, eventually causing him terrible

* The reflecting circle—an instrument based on the same principles as the sextant but with a completely circular arc.

pain. Although he lived just long enough to see the printed atlas of the charts on which he had lavished so much care,[25] he was already unconscious when the bound volumes of his narrative arrived. It is said that Ann put his hand on them, but Flinders died the next day, on July 19, 1814. He was forty. No trace of his grave in Hampstead remains.[26] The books did not sell well and the Admiralty refused to grant Ann a pension (as they had done for Cook's widow), so she and her daughter faced great hardship. At last in 1853 the state governments of New South Wales and Victoria offered help, but Ann had already died.[27] Happily, however, these pensions devolved on her daughter, whose son, Sir William Flinders-Petrie, seems to have inherited some of his grandfather's skills: his meticulous surveys of Stonehenge and the Egyptian pyramids were to launch his illustrious career as an archaeologist.*

Though Flinders himself was too modest to attach his own name to any of his discoveries (he named Flinders Island after his brother), he has certainly not been forgotten in Australia, where many features of the landscape and coastline bear his name, such as the Flinders Mountain Ranges in South Australia, the Flinders Passage on the Great Barrier Reef, and the Flinders Islands off the coast of northern Queensland. Some of his charts remained in print until 1912.

Flinders's personal papers and correspondence reveal far more clearly than his official account what a hard and lonely life he led as commander of the *Investigator*, responsible not only for the efficient discharge of his onerous surveying duties, but also for the safety and health of his crew. Nevertheless, at one point in his narrative, Flinders—perhaps unconsciously—lets his

---

* It was Sir William who, in 1918, secured many of the papers left by Matthew Flinders after his death by donating them to the Mitchell Library in New South Wales—on condition that a statue was erected in his grandfather's honor. This was unveiled in Sydney in 1925.

guard drop. Having listed the various seeds that the sailors had planted on Wreck Reef, he regrets their lack of "cocoa nuts," observing that these trees would not only provide a useful beacon to mariners, but also "salutary nourishment" to any future victims of shipwreck. Indeed, Flinders suggests that the planting of coconut trees "ought to be a leading article in the instructions for any succeeding voyage of discovery," and to those who might question the importance of such a plan, he responded that it was from suffering ourselves that "we learn to appreciate the misfortunes and wants of others." Flinders then revealingly quotes a passage (loosely translated) from a novella by Chateaubriand* that he must have read while in Mauritius:

> *The human heart . . . resembles certain medicinal trees which yield not their healing balm until they have themselves been wounded.*[28]

His beloved cat, Trim, though never mentioned in Flinders's formal record, was a constant comfort and shared his bed aboard the *Investigator*:

> *Trim took a fancy to nautical astronomy. When an officer took lunar or other observations, he would place himself by the time-keeper, and consider the motion of the hands, and apparently the uses of the instrument, with much earnest attention; he would try to touch the second hand, listen to the ticking, and walk all around the piece to assure himself whether or no it might not be a living animal. And mewing to the young gentleman whose business it was to mark down the time, seemed to ask an explanation.*

---

* *Atala*, first published in 1801, was inspired by Chateaubriand's travels in North America.

When the officer had made his observation, the cry of Stop! roused Trim from his meditation; he cocked his tail, and running up the rigging near to the officer, mewed to know the meaning of all those proceedings. Finding at length that nature had not designed him for an astronomer, Trim had too much good sense to continue a useless pursuit; but a musket ball slung with a piece of twine, and made to whirl round upon the deck by a slight motion of the finger, never failed to attract his notice, and to give him pleasure; perhaps from bearing a near resemblance to the movement of his favourite planet the moon, in her orbit around the primary which we inhabit.[29]

Trim was presumably named after Corporal Trim, the devoted follower of Uncle Toby, in Laurence Sterne's marvelously strange and touching novel *Tristram Shandy*. So we may reasonably suppose that Flinders had a soft spot for Uncle Toby, perhaps the sweetest-natured figure in English literature—the original "man who would not hurt a fly."[30] Sadly, Trim disappeared while his master was held on the Isle de France. Flinders's whimsical tribute to his feline companion is very moving, especially bearing in mind how little time he himself had yet to live:

To the memory of Trim, the best and most illustrious of his Race,—the most affectionate of friends,—faithful of servants, and best of creatures.

He made the Tour of the Globe, and a voyage to Australia, which he circumnavigated; and was ever the delight and pleasure of his fellow voyagers. Returning to Europe in 1803, he was shipwrecked in the Great Equinoxial Ocean; This danger escaped, he sought refuge and assistance at*

---

* That is, equatorial.

the Isle of France, where he was made prisoner, contrary to the laws of Justice, of Humanity, and of French National Faith; and where, alas! he terminated his useful career, by an untimely death, being devoured by the Catophagi of that island. Many a time have I beheld his little merriments with delight, and his superior intelligence with surprise: Never will his like be seen again!

Trim was born in the Southern Indian Ocean, in the year 1799, and perished as above at the Isle of France in 1804.

Peace be to his shade, and Honour to his memory.[31]

Chapter 14

Voyages of the *Beagle*

Day 15: Took some sights at 0830. Seas still quite high but blue sky spreading from N. Slept for a bit then had fried eggs for breakfast at 1030. It's amazing how well the eggs are keeping.

By midday it was quite sunny and we took more fixes at 1300 giving noon position of 44°44' N, 27°52' W, 119 miles run. Passing over the Mid-Atlantic Ridge north of the Azores.

During afternoon we passed through a great flock of shearwaters wheeling and diving, and saw more dolphins in the distance. Presumably they were all after a shoal of fish. Lovely rainbows—N force 5–6. No. 2 stays'l and reefed main.

Steering 080° for Land's End. High spirits all round as England gets nearer and BBC radio now clear at night. Celebrated halfway with the tinned chicken, fried, with tinned new potatoes and beans, followed by rice pudding. All cooked by Alexa, who called it "Poulet Atlantique." Sang songs to accompaniment of Colin's "squeeze box" and Alexa's guitar.

The very first European to see the Pacific Ocean was the Spanish soldier Vasco Núñez de Balboa, not, as the poet Keats would have it, "stout Cortez."[1] Nevertheless, when they struggled through the jungle across the Isthmus of Panama in 1513, Balboa and his men may well have looked at each other "with a wild surmise" when that vast blue expanse opened up before them. With the arrogance of the conquistador, Balboa waded into the ocean and claimed all the land within and around it for the king of Spain. Surely the "Spice Islands" with all their riches lay beyond the horizon, as well as China, but how could they be reached? The search for a passage from the Atlantic to the Pacific was soon begun. The Straits of Magellan were discovered by the Portuguese navigator of that name* in 1520 when in command of a Spanish expedition, but for a long time it remained unclear how much farther to the south the American landmass extended. Was it perhaps joined to the great southern continent? In 1578 Francis Drake landed on what seemed to him to be the southernmost point of South America, having been driven to the southeast by storms after entering the Pacific through the Straits of Magellan. However, it was only in 1616 that a Dutch expedition led by Isaac Le Maire and Willem Corneliszoon Schouten—backed by the burghers of Hoorn—entered the Pacific from the Atlantic by the open sea. It was they who gave Cape Horn its name.

Bougainville had not enjoyed his long and arduous passage through the Straits of Magellan in 1767–68, and his experience was typical. Baffling headwinds, dangerously violent downdrafts from the surrounding mountains, frequent heavy rain, and low temperatures even in summer made the Straits a most uncomfortable place in the days of sail. However, they did at least offer relatively sheltered waters compared to the wild seas off Cape Horn, especially in winter. The Straits were therefore an important route between the Atlantic and the Pacific and one that badly needed to be properly charted. In the 1820s the Royal Navy, by now committed to the task of systematically surveying the world's oceans, decided to chart not only the Straits themselves, but the many channels and islands of Tierra del Fuego, including Cape Horn—itself a small island. This was one of the

* Actually his name was Fernão de Magalhães, but this was anglicized as Magellan.

most challenging survey missions yet mounted, and it involved a ship that was later to be made world famous by one of its passengers, Charles Darwin.

The *Beagle*, under the command of Lieutenant Pringle Stokes, first sailed for South America in May 1826[2] as tender to the *Adventure*, commanded by Captain Phillip Parker King (1793–1856), who was in overall charge of the survey expedition. As we have already heard, King, as a boy of ten, had witnessed the return of Flinders from Wreck Reef, and—perhaps inspired by this dramatic experience—he later brilliantly completed the survey of the Australian coast that Flinders had begun. Stokes, though a highly experienced sailor, was new to survey work. King was therefore taking a risk by giving him the tough assignment of exploring the exposed western entrance to the Straits of Magellan, where the strong prevailing westerly winds, coupled with the imposing Pacific swells, make the reef-strewn coast a formidably dangerous lee shore.

During that first season, however, Stokes proved more than equal to the task, though the conditions the *Beagle* encountered were testing in the extreme. The *Beagle* was an even smaller vessel than the *Investigator*—only 235 tons, with a complement of about sixty officers and men. Approaching the western extremity of the Straits of Magellan, she encountered a heavy breaking sea caused by the deep swell of the Pacific, and Stokes found an anchorage for the night under Cape Tamar. The following evening they nearly reached another, farther to the west, under Cape Phillip, but the weather was atrocious:

> *About seven in the evening* [recorded Stokes] *we were assailed by a squall, which burst upon the ship with a fury far surpassing all that preceded it: had not sail been shortened in time, not a stick would have been left standing, or she must have capsized. As it was, the squall hove her so much over on her broadside, that the boat which was*

hanging at the starboard quarter was washed away. . . . On closing [Cape Tamar], the weather became so thick that at times we could scarcely see two ships' lengths a-head.[3]

Stokes dryly commented that "these circumstances were not in favour of exploring unknown bays," and he therefore reluctantly decided to forfeit the progress they had made to windward, and headed back to the anchorage they had earlier left. As King later explained:

Even this was a dangerous attempt; they could hardly discern any part of the high land, and when before the wind, could not avoid the ship's going much too fast. While running about eight knots, a violent shock—a lift forward—heel over—and downward plunge—electrified everyone; but before they could look round, she was scudding along, as before, having fairly leaped over the rock.[4]

Stokes now decided to leave the ship in Tamar Bay and gamely undertook a survey mission in the cutter, discovering several well-sheltered anchorages in the process:

Our discomfort in an open boat was very great, since we were all constantly wet to the skin. In trying to double the various headlands, we were repeatedly obliged (after hours of ineffectual struggle against sea and wind) to desist from useless labour, and take refuge in the nearest cove which lay to leeward.[5]

At last the *Beagle* managed to reach the appropriately named Harbor of Mercy, thirty days after leaving Port Famine.* During

---

* So named for the plight of the Spanish settlers who tried and failed to found a colony there in the sixteenth century under the leadership of Pedro Sarmiento de

the next few days, the *Beagle* was employed—to use King's words—"in the most exposed, the least known, and the most dangerous part of the Strait."[6] Stokes, nevertheless, continued to gather all the useful data he could. Incessant rain and thick clouds delayed the observations necessary for making an island just outside the Harbor of Mercy, the southern end of his baseline for the trigonometrical connection of the coasts and islands near the western entrance of this weather-beaten strait. But on February 20, 1827:

> *I weighed and beat to windward, intending to search for anchorage on the north shore, where I might land and fix the northern end of our base line. In the evening we anchored in an archipelago of islands, the real danger of whose vicinity was much increased to the eye by rocks, scattered in every direction, and high breakers, occasioned doubtless by reefs under water. . . . The number and contiguity of the rocks, below as well as above water, render it a most hazardous place for any square-rigged vessel: nothing but the particular duty on which I was ordered would have induced me to venture among them.*[7]

Stokes landed on one of the islands to fix its position. Then "after extricating ourselves from this labyrinth (not without much difficulty and danger), we beat to the westward," hoping to reach Cape Victory, the northwestern limit of the Strait. He was forced back, however, to the Harbor of Mercy, and then, taking advantage of a brief spell of fine weather, set out again in an open boat:

Gamboa, a mathematician and astronomer who had sailed with Mendaña to the Solomon Islands.

> *After pulling, in earnest, for six hours, we landed upon Cape Victory . . . and there, with a sextant, artificial horizon, and chronometer, ascertained the position of this remarkable promontory. From an eminence, eight hundred feet above the sea, we had a commanding view of the adjacent coasts, as well as the vast Pacific, which enabled us to rectify former material errors. Late in the evening we were fortunate enough to get safely on board again, which, considering the usual weather here, and the heavy sea, was unexpected success.*[8]

The passages quoted give some flavor of Stokes's achievements during the 1827 survey season. When the *Adventure* and the *Beagle* returned in 1828 after a break in Montevideo, now accompanied by a schooner called the *Adelaide*, King had no hesitation in sending Stokes and his crew off to survey the coast of southern Chile, farther to the north. It was winter, and they were frequently in danger. In June, at the northernmost limit of their expedition, the *Beagle* was storm-bound in an anchorage on the eastern side of the Gulf of Peñas on the coast of Chile in weather that Stokes described as the worst he had ever experienced:

> *Nothing could be more dreary than the scene around us. The lofty, bleak and barren heights that surround the inhospitable shores of this inlet, were covered, even low down their sides, with dense clouds, upon which the fierce squalls that assailed us beat, without causing any change . . . and, as if to complete the dreariness and utter desolation of the scene, even birds seemed to shun its neighbourhood. The weather was that in which (as Thompson emphatically says) "the soul of man dies within him."*

Stokes commented that they had often been compelled to anchor in places where they were "exposed to great risk and danger" but he judged that their present situation was "by far the most perilous" to which they had yet been exposed. Any sailor will sympathize with this judgment. The *Beagle*'s three anchors could not safely be relied on, as the "terrifically violent" squalls to which they were exposed might pluck them all out of the ground at any moment. Astern of the *Beagle*, at a distance of only three hundred feet, lay rocks and rocky islets "upon which a furious surf raged."[9] If the anchors once started to drag, the ship would be on the rocks in minutes, and the chances of anyone surviving would be negligible. All they could do was to wait.

Eventually the wind eased and the *Beagle* was able to escape her terrifying predicament. The surgeon then advised Stokes that the "long-continued succession of incessant and heavy rain, accompanied by strong gales" was seriously affecting the health of the crew and that if the many who were suffering from respiratory and rheumatic complaints were to get well, they needed to have some rest.[10] Stokes agreed to this proposal, but it was not just the crew who were suffering. The gruelling mission was proving too much for him. He shut himself in his cabin, becoming listless and inattentive to what was going on around him. His choice of the quotation from *The Seasons* by the Scottish poet James Thomson (1700–1748)—not Thompson, as Stokes gives it—is all too revealing of his state of mind, especially if read in its context:

Thus Winter falls,
A heavy gloom oppressive o'er the world,
Through Nature shedding influence malign,
And rouses up the seeds of dark disease.
The soul of man dies in him, loathing life,
And black with more than melancholy views.

By the time the ship returned to Port Famine (in the middle reaches of the Straits of Magellan), where the *Adventure* was awaiting her, their food supply was exhausted and Stokes was dangerously depressed. King could see at once that he was not himself, though he did not fully appreciate the seriousness of his condition. On August 1, 1828, poor Stokes shot himself in the head, though the wound was not immediately fatal. "During the delirium that ensued," King reported, "his mind wandered to many of the circumstances, and hair-breadth escapes, of the Beagle's cruise." After rallying briefly, and "lingering in most intense pain," Stokes died on August 12. King wrote that the "severe hardships of the cruise, the dreadful weather experienced, and the dangerous situations in which they were so constantly exposed—caused, as I was afterwards informed, such intense anxiety in his excitable mind, that it became at times so disordered, as to cause the greatest apprehension for the consequences."[11]

The next commander of the *Beagle* was a remarkable young officer named Robert FitzRoy. Born in 1805, he was a descendant of King Charles II, and his uncle Lord Castlereagh had served as foreign secretary during and after the Napoleonic Wars. But FitzRoy was not the sort of man to rely on influence to advance his career. He was highly intelligent, extremely hardworking, and very serious-minded. He passed the Royal Naval College in 1825 with full marks in the final examination (the first person ever to do so), which guaranteed his immediate promotion to the rank of lieutenant at the early age of nineteen. By this time FitzRoy already had five years' experience at sea, including cruises in the Mediterranean and South America. After serving as first officer of the frigate *Thetis* he was given the coveted post of flag officer to Admiral Sir Robert Otway aboard the *Ganges*, based in Rio de Janeiro. Despite being only twenty-three and having no previous survey experience, he was appointed by Otway in October 1828 to supersede one of the *Beagle*'s own officers (Skyring), who had been given an acting promotion following the death of Stokes.

This was an awkward beginning, but FitzRoy—though touchy and hot-tempered—was a quick learner, physically tough, very determined, and a natural leader who commanded instant respect from his officers and crew. During the next two years they made a series of significant discoveries among the complex network of islands and channels north and south of the Straits of Magellan, as well as exploring much of the seaward coast of Tierra del Fuego. The conditions they encountered were very demanding, but FitzRoy relished the challenges he faced and even found time to admire the scenery as the *Beagle* headed west toward Lyell Sound on April 19, 1829:

> *The night was one of the most beautiful I have ever seen; nearly calm, the sky clear of clouds, excepting a few large white masses, which at times passed over the bright full moon: whose light striking upon the snow-covered summits of the mountains by which we were surrounded, contrasted strongly with their dark gloomy bases, and gave an effect to the scene which I shall never forget.*[12]

Later, as he headed into the unknown waters on the north side of the Straits of Magellan in a twenty-four-foot open boat, he complained that "some voyagers" had exaggerated the grimness of these coasts:

> *it is true that the peaks of the mountains are covered with snow, and those sides exposed to the prevailing west winds are barren, and rugged; but every sheltered spot is covered with vegetation, and large trees seem to grow almost upon the bare rock. I was strongly reminded of some of the Greek*

islands in winter, when they also have a share of snow on their mountains.[13]

FitzRoy and his men had the advantage of the newly invented tinned meats, but their meals might not have satisfied a gourmet, and the southern winter was closing in:

After dark, we returned to the cutter and partook of a large mess, made of the swan we had shot, the coots, some limpets, and preserved meat. The shortness of the days was becoming very inconvenient . . . but some of the nights were so fine, that I got many sets of observations of the moon and stars.[14]

There were, of course, moments of great danger. Returning in the cutter from Skyring Water (named after the very able lieutenant whom FitzRoy had superseded), they were caught in a gale on a lee shore in Otway Water in freezing temperatures. After taking in reef after reef, they were at last forced to drop the sail and row. The shore, on which a high surf was breaking, offered no prospect of shelter and to have attempted to land would have been "folly." At three in the afternoon they were embayed and unable to get clear except by rowing even harder. FitzRoy's boat was deeply laden, and as their clothes and bags got soaked, progress became more and more difficult. They threw a bag of fuel overboard, but kept everything else to the last. At sunset the sea was higher, and the wind as strong as ever:

Night, and having hung on our oars five hours, made me think of beaching the boat to save the men; for in a sea so short and breaking, it was not likely she would live much longer. At any time in the afternoon, momentary neglect, allowing a wave to take her improperly, would have swamped us; and after dark it was worse. Shortly after bearing up, a heavy sea broke over my back [*FitzRoy must have been rowing in the bows*], *and half filled the boat: we were baling away, expecting its successor, and had little thoughts of the boat living, when—quite suddenly—the sea fell, and soon after the wind became moderate. So extraordinary was the change, that the men, by one impulse, lay on their oars, and looked about to see what had happened. Probably we had passed the place where the tide was setting against the wind. . . . About an hour after midnight, we landed in safety at Donkin Cove; so tired, and numbed by the cold, for it was freezing sharply, that we could hardly get out of the boat.*[15]

At last, after an absence of six weeks in an open boat, in the depths of winter, FitzRoy and his crew returned safely to the *Beagle* on June 8:

I never was fully aware of the comfort of a bed until this night. Not even a frost-bitten foot could prevent me from sleeping soundly for the first time during many nights.[16]

In late November 1829, on their return from the island of Chiloé, the *Beagle* began her examination of the southernmost part of the coast from Cape Pillar all the way round Cape Horn itself. This was an area that Cook had touched on in December 1774, but his chart consisted largely of dotted lines. FitzRoy was determined to obtain as much detailed information as possible about this wild and desolate shoreline, but it was impossible to conduct a running survey:*

---

* Parts of the exposed southwestern coast of Tierra del Fuego remain uncharted to this day.

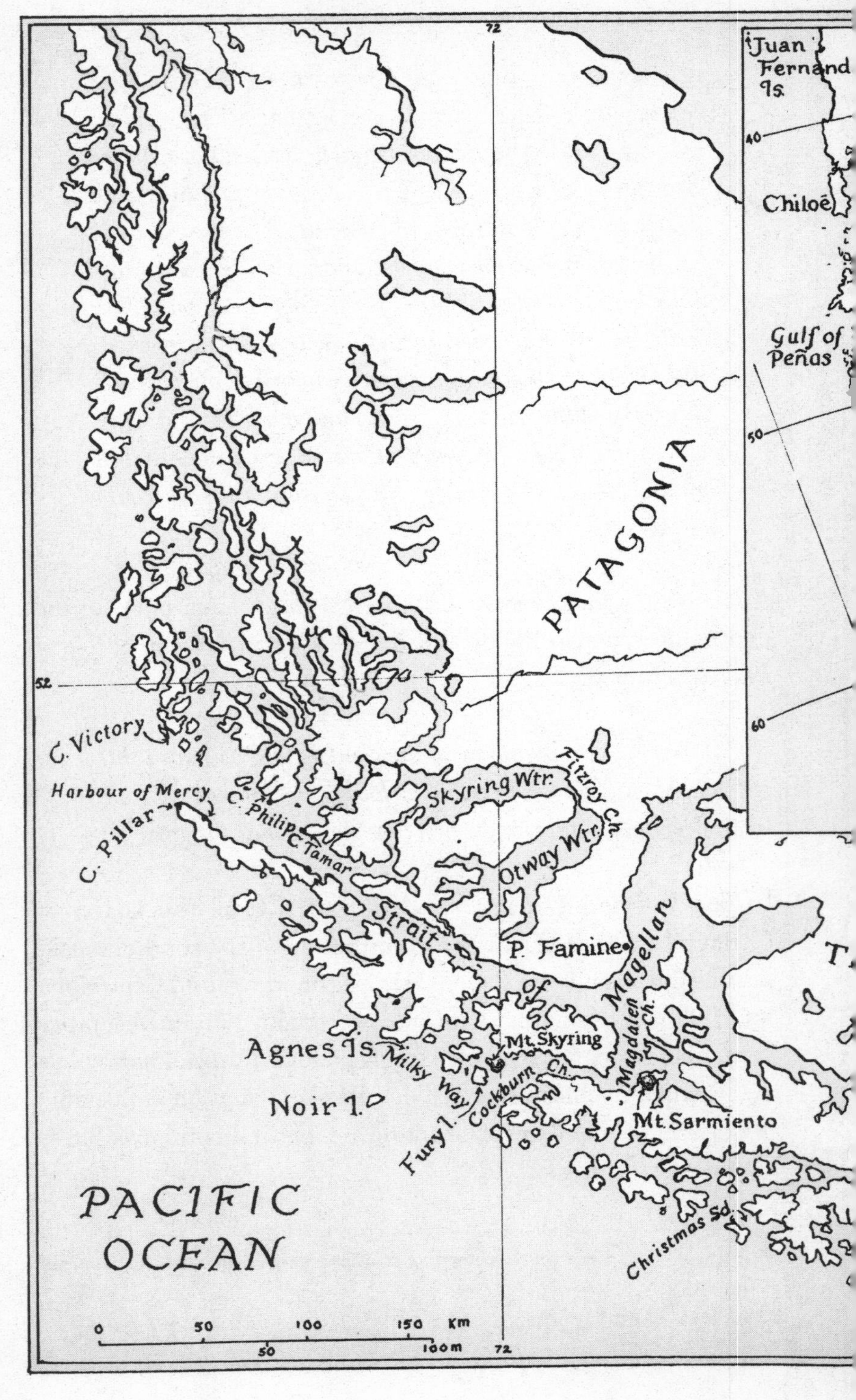

72
Juan Fernand Is.
40
Chiloē
Gulf of Peñas
50
60
PATAGONIA
52
C. Victory
Harbour of Mercy
C. Pillar
C. Philip
C. Tamar
Skyring Wtr.
Fitzroy Ch.
Otway Wtr.
Strait
of
Magellan
P. Famine
Magdalen Ch.
Mt. Skyring
Agnes Is.
Milky Way
Cockburn Ch.
Noir I.
Fury I.
Mt. Sarmiento
Christmas Sd.
PACIFIC OCEAN
0
50
100
150
Km
50
100m
72

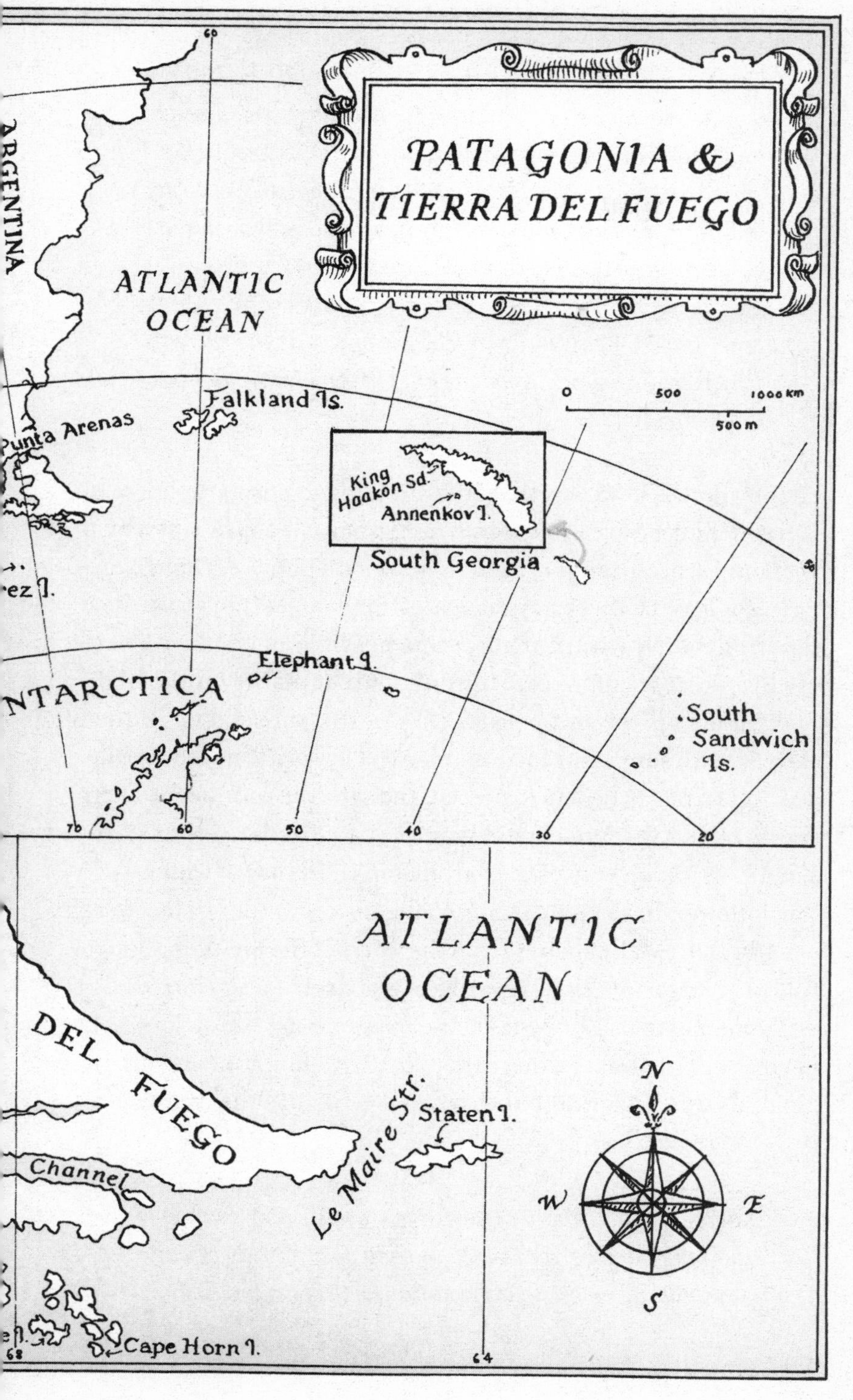
PATAGONIA & TIERRA DEL FUEGO
ARGENTINA
ATLANTIC OCEAN
Falkland Is.
Punta Arenas
0
500
1000 km
500 m
King Haakon Sd.
Annenkov I.
South Georgia
Elephant I.
NTARCTICA
South Sandwich Is.
70
60
50
40
30
20
ATLANTIC OCEAN
DEL FUEGO
Channel
Le Maire Str.
Staten I.
N
W
E
S
Cape Horn I.
68
64

> *On that coast the weather was so continually bad, there was so much swell, and the water near the steep precipitous shores always so deep, that anchorage (except in harbours) was impracticable: boats were seldom able to assist (while under way), and the bearing compass, though particularly good, and well placed, was of very little use: it was therefore never trusted for important bearings. Another impediment, and not a slight one, was the current: which set irregularly from one knot to three knots an hour, along the shore.*[17]

FitzRoy noted, however, that "the stormy and desolate shores of Tierra del Fuego are broken into numerous islands, about which anchorages are abundant, and they are excellent." Although entering or leaving these havens was difficult and often risky, once safely anchored within them it was possible to conduct a survey by taking accurate bearings of distinct marks on the high, rocky shore or "the sharp peaks of more distant heights." He declared—with some bravado—that the coast was not nearly as rugged and harsh as he had expected, though "the number of islets and breakers is quite enough to give it a most dangerous character."[18] With the help of the barometer, in the use of which FitzRoy followed in the pioneering footsteps of Flinders, they were usually able to take shelter from the worst of the weather, though finding reasonably safe anchorages was itself a risky business.

To the northeast of Noir Island, where Anson had so nearly been wrecked almost ninety years earlier, and in the neighborhood of Cape Kempe and the Agnes Islands, lay the "many perils" of "Breaker Bay":

> *Having approached as near as we could, and sounded, and taken angles, we steered so as to pass outside of some very outlying rocks, near the middle of the bay; for in-shore*

> *of them, I saw from the mast-head numerous breakers, rocks and islets, in every direction. A worse place for a ship could scarcely be found; for, supposing thick weather to come on when in the depth of the bay, she would have lurking rocks and islets just awash with the water, on all sides of her, and no guide to take her clear of them, for soundings would be useless.*
>
> *. . . the chart of it, with all its stars to mark the rocks, looks like a map of the heavens, rather than part of the earth.*[19]

This fearsome stretch of water matches the description of "the Milky Way" on which Joshua Slocum's *Spray* came close to being lost seventy-odd years later.

Once the *Beagle* was safely at anchor, shore parties climbed the surrounding hills and mountains—often with great difficulty—to obtain theodolite bearings, while the ship's boats explored the complex networks of islands, taking soundings, sextant angles, and compass bearings (though local magnetic anomalies sometimes made the latter problematic). The crews of the boats faced many dangers and privations, too, not least because in some places the natives were hostile. Whenever the weather permitted, sextant sights were taken for latitude and to check the chronometers. Gradually FitzRoy was able to connect his new chain of observations with those made by his colleagues farther to the north, thereby extending the reach of the survey to embrace many prominent features of Tierra del Fuego. A climb to the top of Mount Skyring proved especially useful:

> *as the day was perfectly clear, and free from clouds, every point of land was visible, which can at any time be seen from that summit. Mount Sarmiento appeared in all its grandeur, towering above the other mountains to at least*

> *twice their height, and entirely covered with snow. Having set the theodolite to a painted post, fixed on shore near the Beagle (five miles distant), from which I had previously obtained the exact astronomical bearing of the spot on which the theodolite was placed; I obtained a most satisfactory round of angles, including most of the remarkable peaks, islands, and capes, within a range of forty miles from this mountain.*[20]

The theft by natives of a crucially important whaleboat caused a great deal of trouble. Her crew found themselves marooned in the freezing cold and with no means of returning to the *Beagle*. In desperation they enterprisingly constructed a coracle out of sticks woven together, lined with canvas and daubed with mud. In this frail conveyance two men somehow managed to get back to the ship after paddling for thirty-six hours through heavy seas. FitzRoy then rescued the remaining members of the boat's crew and spent weeks trying to recover the missing craft. There is something almost manic about the fury and determination with which he pursued this goal, and there is a hint here of the mental illness that was later to trouble him so deeply. One native was killed in a skirmish, and FitzRoy ruthlessly kidnapped three children to use as hostages. But it was all in vain. In the end, they found a quiet anchorage in Christmas Sound (so named by Cook), where the carpenter set about building a new whaler, while the remaining boats continued their survey work—discovering, without fully exploring, an important new passage later named the Beagle Channel.

On April 18, 1830, the *Beagle* reached St. Martin's Cove on Hermite Island, just to the northwest of Cape Horn. The weather was fair, and after a preliminary reconnaissance FitzRoy set off in a boat with five days' provisions, a good chronometer, and all the other instruments—including "two good sextants"—to

Horn Island itself. They pitched camp for the night, and the next day left a "memorial" of their visit to the southern tip of South America in a stone jar:

> *At daybreak [on April 20] we commenced our walk across the island, each carrying his load; and by the time the sun was high enough for observing, we were near the summit, and exactly in its meridian; so we stopped while I took two sets of sights and a round of angles. Soon afterwards we reached the highest point of the Cape, and immediately began our work; I and my coxswain, with the instruments; and Lieut. Kempe with the boat's crew raising a pile of stones over the memorial. . . . We drank the health of His Majesty King George the Fourth, and gave three hearty cheers, standing round the Union Jack.*[21]

The *Beagle* then sailed out to the Diego Ramírez Islands, which had been discovered in 1619 by a Spanish expedition and named after the navigator, Diego Ramírez de Arellano. They lie menacingly almost sixty nautical miles to the southwest of Cape Horn: "The two largest are about two hundred feet high . . . there is a shingle beach on one . . . where a boat may be hauled up in safety. . . . A furious surf breaks against the west shore, and sends spray over the whole island. There is no sheltered anchorage for a vessel."[22] Having fixed their position, FitzRoy returned to the mainland before rounding Cape Horn in "beautifully fine weather, more like the climate of Madeira than that of fifty-six south latitude."[23] He closely examined the coast to the northeast as far as the Le Maire Strait and Staten Island, as well as the eastern part of the Beagle Channel, and then the *Beagle* headed north, with three young Tierra del Fuegians aboard. FitzRoy had—quixotically—decided to take them to England in the hope of educating them before returning them to their home, where they might, he fondly hoped, civilize their compatriots. The *Beagle* caught up with the *Adventure* in Rio de Janeiro and the two ships sailed together for England on August 6, arriving in Plymouth on October 14, 1830, "after a most tedious passage."

FITZROY HAD MADE his mark, but this was just the first phase of his career as a naval hydrographer. In 1831, at age twenty-six, he was ordered by the Admiralty to survey the parts of the coasts of Argentina, the Falkland Islands, Chile, and Peru that remained uncharted, and then to return home via the Cape of Good Hope. An additional purpose was to return the Tierra del Fuegians, who had by now received an education in Walthamstow (then a village on the northern edge of London) at FitzRoy's expense.* He invited a young naturalist named Charles Darwin to accompany him on the voyage of circumnavigation. FitzRoy's aim in doing so was partly to ensure that the many opportunities for scientific discovery were put to good use and partly to alleviate the loneliness of command: he may already have been aware of his own depressive tendencies, and the miserable fate of Stokes would surely have served as a warning to him. Darwin described FitzRoy as his "Beau Ideal" of a captain, and greatly admired his dedication and professionalism. He wrote to his sister that he had never before come across a man he "could fancy being a Napoleon or a Nelson":

> *I should not call him clever, yet I feel convinced nothing is too great or too high for him. His ascendancy over everybody is quite curious: the extent to which every officer and*

* FitzRoy duly established the three young people, with every modern convenience, on the southern shores of Tierra del Fuego, but the experiment was a complete failure. They soon fell out with each other, abandoned the habits they had learned in London, and reverted to their former way of life.

> *man feels the slightest rebuke or praise, would have been before seeing him, incomprehensible.*[24]

FitzRoy's instructions run to almost twenty pages in the published account of his voyage, amounting to a formidable list of requirements. In addition to the extensive but—by comparison with his previous experience—largely routine survey work, the commander was to obtain "a series of well-selected meridian distances [that is, differences in longitude] in traversing the Pacific Ocean." The *Beagle*'s famous voyage embraced the Galapagos Islands, Tahiti, New Zealand, Australia, the Cocos-Keeling Islands, Mauritius, the Cape of Good Hope, as well as various Atlantic islands, including the Azores—and FitzRoy discovered a few new Pacific islands on the way. One of his very first tasks, however, was to determine the exact longitude of Rio de Janeiro, which even at this date—more than three hundred years after the Portuguese had first visited it—was still a matter of dispute:

> *A considerable difference still exists* [the orders state] *in the longitude of Rio de Janeiro, as determined by Captains King, Beechey, and Foster, on the one hand, and Captain W. F. Owen, Baron Roussin, and the Portuguese astronomers, on the other; and as all our meridian distances in South America are measured from thence, it becomes a matter of importance to decide between these conflicting authorities. Few vessels will have ever left this country with a better set of chronometers, both public and private, than the Beagle; and if her voyage be made in short stages, in order to detect the changes which take place in all chronometers during a continuous increase of temperature, it will probably enable us to reduce that difference within limits too small to be of much import in our future conclusions.*[25]

FitzRoy brought no fewer than twenty-two chronometers with him—many of which he paid for himself—as well as an instrument maker whose sole function was to maintain them. In keeping with well-established practice, however, he methodically checked their performance at every opportunity by astronomical means. By the time the *Beagle* returned home in October 1836, some of the watches had stopped, others had exhibited sudden changes of rate, and the mainspring of one had broken when it had been "going admirably."[26] Only half of the original twenty-two were still in good order. So even at this stage in the development of the watchmaker's art the chronometer was not an instrument on which a ship's captain could rely with complete confidence.

During the four-year voyage of the *Beagle,* FitzRoy and his colleagues gathered abundant information about tides, ocean currents, weather patterns, earthquakes, mountain ranges, isolated reefs, fossils, animal and plant species, the structure of coral atolls, and much else besides—including the customs of the many different native populations they encountered. These discoveries, in Darwin's hands, were to inspire the epoch-making theory of evolution by natural selection and gravely undermine traditional religious beliefs—greatly to FitzRoy's dismay. From the point of view of the Admiralty, however, the most valuable product of the voyage was the mass of navigational data that FitzRoy had gathered with the help of his sextants and chronometers.

For all the technological progress that had been made, navigation still remained a perilous business—as an episode early on in the *Beagle*'s circumnavigation makes clear. While on the coast of Brazil, FitzRoy made a point of investigating the disturbing loss of the crack frigate *Thetis* (in which he had earlier served) on the bold headland of Cape Frio, about seventy miles east of Rio de Janeiro. The court-martial proceedings, which FitzRoy quotes in his journal, give a dramatic account of the wreck. The *Thetis* had set sail from Rio in foggy weather on December 4, 1830, first heading out to sea and then shaping a generally eastward course parallel with the coast. By 4 P.M. the following day, relying on DR "as neither sun, moon or stars had been seen," it was reckoned that the ship was twenty-four miles off the Cape and her course was altered accordingly to northeast by east. After a brief clearing when—reassuringly—no land could be seen, "thick and rainy" weather set in and by 8 P.M. "nothing could be distinguished half a ship's length distant":

> *Soon after eight one of the look-out men . . . said to another man on the forecastle, "Look how fast that squall is coming" . . . and next moment, "Land a-head," "Hard a-port," rung in the ears of the startled crew, and were echoed terribly by the crashing bowsprit, and thundering fall of the ponderous masts.*

The hull did not immediately strike the rocks, as the ship answered the helm very quickly, but as she turned the bowsprit broke and the lee yardarm irons "struck fire from the rocky precipice" as they grated harshly against it, while the boom ends snapped off "like icicles."

> *All three masts fell . . . strewing the deck with killed and wounded men. An immense black barrier impended horribly, against which breakers were dashing with an ominous sound; but the ship's hull was still uninjured. Sentries were placed over the spirit-room; a sail was hoisted upon the stump of the main-mast; the winches were manned; guns fired; rockets sent up, and blue-lights [fireworks used as distress signals] burned. . . .*[27]

The boats were cleared away for lowering and an anchor was let go, but the water was so deep that before the anchor could grip the ship's stern drifted onto a shelving rock. Some of the crew then tried to jump ashore, but many slipped and were drowned in the surf, or crushed against the rocks. The ship was now starting to sink and successive waves threw her stern on the rocks. Luckily the battered ship was then driven into a small cove, and by daylight all the survivors were saved by being hauled through the surf along a rope stretched to the shore. Many, however, were terribly bruised and lacerated, and twenty-five had lost their lives. Some of the officers obtained horses, and a guide took them back to Rio de Janeiro, where "the melancholy news was communicated to the commander-in-chief."

The loss of such a large quantity of treasure, as well as a fine ship and many men, no doubt added to the sense of dismay evoked by this disaster. There could hardly be a clearer demonstration of the unreliability of DR, but FitzRoy was anxious to establish how such a serious navigational error could have occurred within just nineteen hours of the *Thetis* setting sail from Rio. Skeptical of the theory that the accident had been caused by a local magnetic disturbance affecting the steering compass, he concluded that the *Thetis* had been pushed off course by a strong but hitherto unknown seasonal current setting the ship to the north. His conclusion was generous, and wry:

> *If a man of war is accidentally lost, a degree of astonishment is expressed at the unexpected fate of a fine ship, well found, well manned, and well officered; and blame is imputed to some one: but before admitting a hastily-formed opinion as fact, much inquiry is necessary.*
>
> *. . . Those who never run any risk; who sail only when the wind is fair; who heave to when approaching land, though perhaps a day's sail distant; and who even delay*

> *the performance of urgent duties until they can be done easily and quite safely; are, doubtless, extremely prudent persons:—but rather unlike those officers whose names will never be forgotten while England has a navy.*[28]

Even if FitzRoy had died in 1836, his own name would deserve a place in that illustrious list. But a varied and distinguished career still lay ahead of him. Soon after his return, he married and published his account of the *Beagle* voyages. Having been elected a member of Parliament in 1841, he turned his back on a potentially successful political career by accepting a posting as governor of New Zealand in 1844. The infant British colony was then in the throes of bitter disputes both among the European settlers and between them and the native Maori population. FitzRoy lacked the necessary resources, both financial and military, to manage the situation successfully, but his conciliatory approach to the Maoris—born of his increasingly fervent Christian beliefs—was exemplary. Party political maneuvering in London was largely responsible for his early recall in 1845, after which he returned to active service as commander of the Royal Navy's first propeller-driven frigate. But having suffered a crippling bout of depression, brought on in part by growing financial problems, FitzRoy resigned his command in 1850. He was never offered another.

Some of FitzRoy's greatest achievements came at the end of his career. Building on his own experience and detailed observations at sea, and those of other scientifically minded mariners like Flinders, he played an important role in the development of meteorological science. In 1854 he became the first head of what is now known as the Meteorological Office. In that role, and in the teeth of bureaucratic obstruction and ill-informed public criticism, he established in 1861 a pioneering system of "storm warnings" based on data gathered telegraphically from observation stations around the United Kingdom. He invented the visual signals ("storm cones") that were hoisted at the entrance of ports to warn ships of approaching heavy weather, and they were soon replicated in other countries. Though briefly suspended after his death, this "storm warning" system remained in place for many years. He also designed a robust barometer that was issued as standard to ships, with an instruction manual that would enable the commander to foretell likely changes in the weather.[29] FitzRoy went on to develop the first "weather forecasts"—a term that he himself coined—and in 1862 published a popular work on meteorology, *The Weather Book*. Though some of the ideas he put forward have not stood the test of time, his researches enabled him to recognize the existence of ridges and troughs, as well as wider areas of low and high atmospheric pressure, all key determinants of the weather.[30]

An episode recounted in *The Weather Book* dramatically illustrated the value of barometric readings at sea. On FitzRoy's return voyage from New Zealand in 1846, the ship in which he and his family were sailing anchored at the western end of the Straits of Magellan—a part of the world that he knew better than most. The weather was fair and the ship's captain seemed content to spend the night riding to a single light anchor, but FitzRoy, having noticed that his two personal barometers were dropping sharply, remonstrated with him. The captain agreed to replace the light anchor with a heavier one and to reduce the ship's windage by striking some of her lighter spars down to the deck, but the barometers continued to fall and, the captain now being peacefully asleep in his bunk, FitzRoy persuaded "a good officer and a few willing men" to drop a second anchor, even though the night as yet remained calm. At 2 A.M. the storm he had predicted hit the ship and, despite all the precautions, she dragged her anchors to "within a stone's throw of sharp granite rocks astern, at some distance from land, near the most exposed outer point of the harbour." But for FitzRoy's intervention she would almost certainly have been lost.[31]

The publication in 1859 of Darwin's revolutionary work *On the Origin of Species* deeply disturbed FitzRoy, who was by now convinced of the literal truth of the Genesis story. He felt personally responsible for the shocking ideas expressed in the new book because Darwin had been sailing with him when carrying out the research that inspired them. His few attempts to combat Darwin's ideas were, as he himself realized, quite futile. Though he reached the rank of vice admiral in 1863, he was by then worn-out by a lifetime of hard work. Depression once again gripped him. Dismayed by the apparently unstoppable progress of Darwinism, worried by his worsening financial problems, and hurt by criticism (much of it unfair) of his meteorological endeavors, in 1865 he cut his throat—just as his uncle Lord Castlereagh had done in 1822.

It was a heartbreaking end to a remarkable career. The Hydrographer of the Navy, George Richards, wrote to FitzRoy's widow that "No naval officer ever did more for the practical benefit of navigation and commerce than he did, and did it too with a means and at an expense to the country that would now be deemed totally inadequate. . . . His works are his best as they will be his most enduring monument, for they will be handed down to generations yet unborn." Despite their differences, even Darwin acknowledged in his autobiography that FitzRoy's character had "many noble features," declaring that "he was devoted to his duty, generous to a fault, bold, determined, indomitably energetic, and an ardent friend to all under his sway."[32]

Though not at first fully appreciated, FitzRoy's contributions to the science and practice of meteorology, and to the welfare of mariners, were at last publicly recognized in 2002 when the BBC Shipping Forecast area Finisterre was renamed "FitzRoy." It is the only sea area named after an individual.

Chapter 15

Slocum Circles the World

Day 16: Wind and seas at last down. N force 3—full main again and No. 1 stays'l making 6 knots.

During morning accompanied again by dolphins. One jumped clear of the water within a few feet and seemed to look right at me. It is very strange how everyone responds to them—we all stood at the bows watching them and laughing with delight. They really seem to be aware of our presence, and they move with such astonishing ease and grace. Perhaps we envy their freedom and their ability to make their home out here where we can live only on sufferance.

Overcast skies and quite cold. Noon position 45°32' N, 25°34' W. Lunch of soup, Ryvita, corned beef and Dundee cake. Colin put a message in an empty whisky bottle giving his address and offering a reward of £5 to the finder for reporting its discovery. Pretty tired, but we're making good progress.

At 1530 the wind backed to NW and eased so we hoisted the big blue genoa again and trickled along at 3–4 knots. More star sights.

Day 17: A good night's rest and up at 0400 to see the dawn and finished Slocum.

Slept after breakfast until Alexa called out that a ship was approaching. She was a smart white general cargo vessel and she came past quite close by on the same course at full speed with the crew lining the rail and waving. Even better Colin managed to make contact on the radio-telephone. The operator was very efficient and helpful giving us the weather forecast (SW 3, 4) *and asking our destination and ETA. He promised to report us to Lloyd's.* She was the "Tanamo"—Dutch and registered in Rotterdam, bound for Cork.*

Combined with suddenly clear skies and a warm sun, this encounter really cheered us. The rest of the day we had a fabulous broad reach at 6½ in S'thly force 3 winds. I even put on my shorts again.

At 1300 I worked out our position on my own again: 46°32' N, 23°57' W. This matched Colin's results pretty closely. Course 090°.

In evening we had cannelloni for dinner and put our watches on an hour. Colin got out his squeeze box and Alexa her guitar and we had another boozy sing-song.

The position-fixing methods employed by FitzRoy differed hardly at all from those first developed in the middle of the eighteenth century, though the accuracy of the ephemeris tables had steadily improved. A year after the *Beagle* returned home in 1836, however, a fortuitous discovery by an American merchant seaman named Sumner[1] led eventually to the development of a completely new approach to celestial navigation.

I first heard the name Sumner when I was helping to deliver a beautiful old yawl called *Wester Till* from Gare Loch on the west coast of Scotland to Brixham on the south coast of England. Our course took us down the Irish Sea and past the Smalls Lighthouse, which marks a dangerous, low-lying reef some thirteen nautical miles off the island of Skomer on the Pembrokeshire coast.† We could see the tall lighthouse clearly, and by measur-

---

* Lloyd's Register of Ships relayed a message from the *Tanamo* to a member of Colin's family, in accordance with instructions left by him before his departure from the United Kingdom. This was a useful service in the days before satellite phones eliminated the need for it.

† Many ships have been lost on the Smalls. In 1991 a diver recovered a brass sword-guard just off the reef—presumably from a wrecked Viking vessel. It dates from *c.* 1100 CE, and each side is finely decorated with a pair of stylized animals, interlaced with thin, snakelike beasts. On the top of the guard two animals with open jaws bite the place where the grip once projected through the guard.

ing its height with the sextant we were able to determine our distance from it and our position. As the lighthouse receded from us, the yawl's owner mentioned that Captain Thomas Sumner had discovered the principle of the "line of equal altitude," on which modern celestial navigation depends, in just this patch of sea. In December 1837 Sumner was in command of a sailing ship bound from Charleston, South Carolina, to the Clyde. The standard method of fixing a ship's position at that date was to deduce the longitude by comparing Greenwich time (recorded by the chronometer) with local time at the ship, and to couple this with the latitude, usually derived from a mer alt. However, the accurate calculation of local time by sextant sight depended on knowing the exact latitude. If for any reason the latitude was unknown, an accurate longitude could be obtained only by taking a sight of the sun when it bore precisely due east or due west, and this might well be impractical.

After passing the meridian of 21 degrees West (several hundred miles out into the Atlantic), Sumner—who was planning to reach the Clyde from the south, via the Irish Sea—was unable to take any sights until the DR suggested that he was within forty miles of the Tuskar Rock lighthouse, off the southeastern tip of Ireland. The wind hauled into the southeast and rose to gale force, while Sumner struggled to keep the ship on the same station until daylight. It was a tense and no doubt chilly night, as his position was uncertain and the lee shore of the Irish coast might be dangerously near. At break of day Sumner would have been relieved to find that no land was in sight, and at 10 A.M. he at last managed to snap a single sight of the sun, noting the time by chronometer. His latitude (52 degrees North) was highly uncertain, as it depended on DR. Using this figure, however, he obtained the local time and calculated that the ship was 15 minutes east of her DR position—nine nautical miles in that latitude. He then tried the calculation again using an assumed latitude

10 minutes farther north, and this placed him 27 miles ENE of the former position. Another calculation based on a latitude 10 minutes still farther to the north yielded a position 54 miles ENE of the first position. These three positions, he soon realized, all lay on a straight line passing through the Smalls Light off the Pembrokeshire coast:

> *It then at once appeared, that the observed altitude must have happened at all the three points, and at the Small's light, and at the ship, at the same instant of time; and it followed that the Small's light must bear E.N.E. if the chronometer was right. Having been convinced of this truth, the ship was kept on her course E.N.E., the wind bearing still S.E., and in less than an hour, Small's light was made, bearing E.N.E. ½ E. and close aboard.*[2]

The straight line formed by the three points was a line of equal altitude. Sumner's account of his accidental discovery appeared in 1843 and was quickly noticed by Royal Navy lieutenant Henry Raper—author of the standard work *The Practice of Navigation*. In an article published a year later, Raper described Sumner's method of deriving a position line from a single sight as "highly ingenious and very useful when a ship is near land."[3] In due course, however, a number of brilliant French sailors and mathematicians were able to develop and systematize Sumner's insight and thereby bring celestial navigation to perfection.

A week out of Halifax I was already fairly proficient at taking a mer alt and calculating our latitude from it. What we really needed, however, was not a line but a point: our actual position in the middle of the ocean. The method that Colin taught me was inspired by the French naval officer Adolphe Marcq St. Hilaire. The grandson of a senior naval officer who had died fighting the

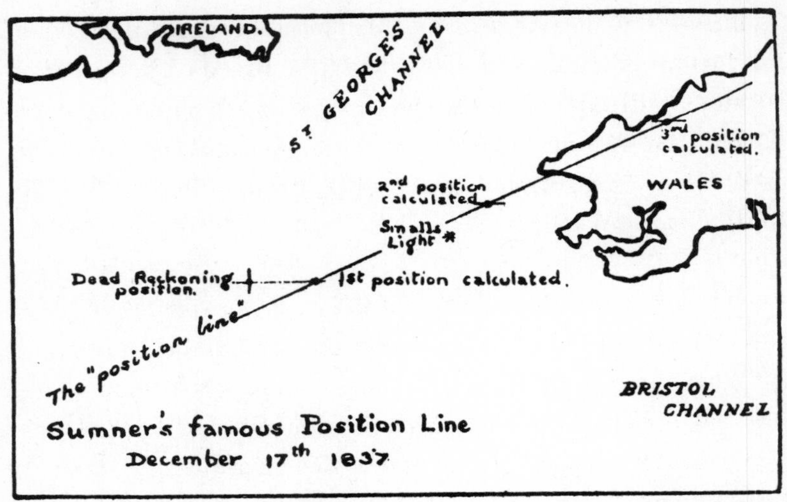

Fig 8: Sumner Line.

British in 1795, Marcq St. Hilaire was born in 1832 and won a scholarship to the Collège Royal in Cherbourg. He then moved on in 1847 to the École Navale (naval college), where the commanding officer recognized his great aptitude. Having joined the navy, he was posted to the Pacific, where he undertook hydrographic work for which he was highly commended.

Marcq St. Hilaire saw service at sea in many parts of the world but in 1870 returned to France in poor health and devoted himself to developing Sumner's position-line method. In 1875, while serving aboard a training ship at Brest, in northwest France, he published a long article describing the new technique that bears his name.[4] Building on the work of a naval colleague, Lieutenant Hilleret, Marcq St. Hilaire devised a novel and elegant solution so radical that he boldly called it the "new navigation."[5]

Imagine taking a sextant sight of a star from a position somewhere in the North Atlantic. If the star is vertically above you, your position must be identical with the star's GP, which is easily determined by reference to the *Nautical Almanac*. So if you know the exact time, you can determine where you are without much trouble. If the star is not overhead, but, say, 45 degrees above the horizon, then you are somewhere on a very large circle with the star's GP at its center—a circle consisting of all the points from which the star's altitude at that moment measures 45 degrees. That is not much help, however, unless you can determine your exact position on that "equal-altitude" circle. Suppose then that you take a sight of another star, in a different part of the sky, at the same time. Once again, you will find yourself on a big circle, but these two equal-altitude circles will intersect—in two places. If you have made your observations correctly, it is simply a matter of deciding which of those two points makes better sense: one of them will be somewhere near your estimated position in the North Atlantic and the other will probably be thousands of miles away. Even if you know only your latitude, the choice should be clear.

In practice, you cannot easily reproduce such large imaginary circles on a paper chart, and if you did, the scale would be so small that it would be hard to determine your position accurately. However, as Marcq St. Hilaire demonstrated, there is no need to draw big circles.

The modern "intercept method" of celestial navigation is based on his insights. Having taken an accurately timed sight, the navigator chooses a place somewhere near to his or her DR position and works out what the height of the observed heavenly body *ought* to be if that "assumed position" were correct at the precise moment the sight was taken, as well as the true bearing (or azimuth) of the observed body's GP from the assumed position. In Marcq St. Hilaire's day it was necessary to calculate these values trigonometrically, but with the introduction of "sight-reduction" tables it became possible simply to look up the answers.* The difference between the precomputed altitude of the observed body shown in the tables and that derived from the *actual* observation (the "intercept"), coupled with the azimuth bearing, permits the navigator to draw a line on the chart: a line of position.[6]

This new graphical system was simpler than any of the older methods. No longer was there any need to calculate the local time and compare it with the time at Greenwich, nor was it necessary to determine the latitude. Moreover, useful sights could be taken at almost any time of day (or night). With the help of sight-reduction tables, you only had to add and subtract accurately and then plot the results on a chart.

Why it should have taken so long for this elegant, simplified technique to emerge is hard to say. Most books on the subject of marine navigation emphasize the great conservatism of sailors. Once they had learned how to do something, and provided it worked, they would doggedly go on doing it that way and would not trouble to seek improvements. There may be some truth in this generalization, but perhaps more important is the fact that few sailors were mathematically sophisticated. For most, celestial navigation was an arcane process that had to be learned by rule, and it would have required a high degree of self-confidence and mathematical skill to challenge the established system. These were qualities that Marcq St. Hilaire plainly had in abundance. He was to reach the rank of admiral but died in 1888 at the early age of fifty-eight. As one authority has written, his name "will always be remembered, so long as mariners practise the marvellous science which enables them to fix their ships with certitude and to rectify their courses with confidence."[7]

---

* Now the process is even easier as handheld computers can perform all the necessary calculations. There is even an iPhone app that works out lunar distances (StarPilot).

UNDER COLIN'S GUIDANCE, I would take a sight of the sun in the morning (timed to the second with the aid of our chronometer) to obtain a single position line that ran roughly from northeast to southwest. A second sun sight in mid-afternoon would yield a position line running from northwest to southeast. Using DR to make allowance for the distance travelled between the two sights, I then "advanced" the first position line on the chart by the right amount and in the right direction: where the two lines crossed was our position. Now we knew both our latitude and our longitude, though neither value had been calculated directly. The same graphical procedure could then be repeated at twilight, making use of one or more of the "navigational" stars, the planets, or even the moon—details of which are listed day by day in the *Nautical Almanac*. By taking a number of sights almost simultaneously, the navigator can generate an immediate fix, in the form of a "cocked hat" of position lines.

Since an error of 1 minute (a mere sixtieth part of a degree) in the observed altitude will necessarily result in an error of one nautical mile in the position line, the whole process depends on the accuracy of the sextant sight itself. This was (and remains) largely a matter of skill and practice, though a good instrument and a clear sky (or at least a gap in the clouds) are obviously essential, as is a clear view of the horizon. In rough conditions—especially in a small vessel like a yacht—it is much harder to take a good sight. Experienced navigators can usually judge whether or not a single sight is good enough to be relied on, but to achieve the highest level of accuracy it is best to take a series of sights and plot them graphically—altitude against time. Deviant sights will then stand out clearly and can be discarded, while the arithmetic mean of the remaining ones is likely to provide a more reliable figure than any single sight.[8]

Accurate timekeeping is also crucial, but by the 1870s this was much less of a challenge than it had been when FitzRoy was serving in the *Beagle* forty years earlier. At that time the Royal Navy was only equipping its ocean-going vessels with "time-keepers"; since then the quality and reliability of marine chronometers had steadily improved and because prices had come down they were being employed increasingly widely. Well-equipped ships in the late nineteenth century usually carried at least three, so that by comparing their rates the navigator might more easily spot the discrepancies of any one instrument. The crucial developments, however, were the arrival of the electric telegraph and the laying of the first successful transoceanic cables in the 1870s and 1880s. Ports all around the world were now able to provide extremely accurate daily time signals (the dropping of a ball or the firing of a gun) for the rating of chronometers. These signals, which had first been introduced during the 1830s at a few major ports where observatories could keep accurate track of the time, offered a much more precise and simple way of checking the time than the old lunar-distance method. And in the early twentieth century, the introduction of radio time signals enabled the navigator to rate his chronometers wherever he happened to find himself—even far out at sea.

The "new navigation" was a major breakthrough and it marked the final stage in the development of celestial navigation. But it was a French discovery, and perhaps partly for that reason the British were slow to adopt it. Captain Lecky—whose wonderfully named *Wrinkles in Practical Navigation* (first published in 1881), with its salty language and charming illustrations, was for many years a bible for professional seamen—was frankly dismissive of the new technique.

He declared himself "averse to the diagram part of the method, which somehow seems . . . out of place in a ship."[9] Eventually, however, its merits became clear even to the most incorrigible Anglo-Saxons, and in the early years of the twentieth century it became the standard method taught to officers in the Royal Navy—among whom Colin was soon to be numbered.[10]

WHAT THEN HAPPENED to lunars? Lecky's comments about lunars in my 1903 edition suggest that by then they were almost obsolete:

> *Once upon a time Lunars used to be the crucial test of a good navigator, but that was . . . when ships were made snug for the night, and the East India "Tea-waggons" took a couple of years to make the round voyage.*
>
> *The writer of these pages, during a long experience at sea in all manner of vessels, from a collier to first-class Royal Mail steamer, has not fallen in with a dozen men who had themselves taken Lunars, or had ever seen others do so. Whether Lunars are worth cultivating or not may, in the minds of some people, still be open to question, but certain it is that they have fallen into disuse—are in fact as dead as Julius Caesar. . . .*[11]

Dead they may have seemed to Lecky, yet some continued to cling to them. Of these perhaps the most famous was Joshua Slocum, whose book I read when sailing across the Atlantic in *Saecwen*. He was the first man to sail around the world single-handed—a feat he performed aboard the *Spray*, an old thirty-five-foot sloop that he had almost completely rebuilt with his own hands.

Slocum was born in 1844 on a small farm in Nova Scotia overlooking the Bay of Fundy and went to sea as a boy. I now realize that we must have passed very close to his birthplace when we sailed in *Saecwen* from Grand Manan Island, as well as following closely in his wake when we later departed from Halifax. Slocum learned his trade in sailing ships, starting as a deckhand and rising to become in 1881 master (and part owner) of "the magnificent ship Northern Light," a large, three-masted clipper, which he proudly described as "the finest American sailing-vessel afloat."[12] This was the high point of his career and his "best command." Slocum, accompanied by his wife and children, had many adventures aboard the *Northern Light*. They rescued Gilbert Islanders adrift in the Pacific,[13] sailed through seas that were literally boiling close to the erupting volcano Krakatoa,[14] and narrowly escaped sinking when the rudder was disabled in a gale off the coast of South Africa.[15] But Slocum lost his command while defending himself in court against charges of cruelly treating an officer who had raised a mutiny against him. Though he was eventually vindicated, it was a murky episode.[16] It did not, however, bring his career to an end: Slocum was to command several other ships before embarking on the retirement project that was to make him world famous.

Slocum set sail in the *Spray* from Gloucester, Massachusetts, in April 1895. He cruised up the coast of Maine to his home town of Westport before calling at Yarmouth on the southwest coast of Nova Scotia, where he laid in water and provisions and "stowed all under deck."

> *At Yarmouth, too, I got my famous tin clock, the only timepiece I carried on the whole voyage. The price of it was a dollar and a half, but on account of the face being smashed the merchant let me have it for a dollar.*[17]

Slocum had left behind his old chronometer—"a good one" that "had been long in disuse"—apparently because it was going to cost fifteen dollars to clean and rate it. "In our newfangled notions of navigation," he complained, "it is supposed that a mariner cannot find his way without one; and I had myself drifted into this way of thinking."[18] But Slocum was to manage just fine without a chronometer. On July 2 the *Spray* sailed from Yarmouth and, after rounding Cape Sable, headed north before taking her departure from George's Island, off Halifax, before dark on the evening of July 3:

> *I watched light after light sink astern as I sailed into the unbounded sea, till Sambro, the last of them all, was below the horizon. The Spray was then alone. . . . I raised the sheen only of the light on the west end of Sable Island, which may also be called the Island of Tragedies. The fog, which till this moment had held off, now lowered over the sea like a pall. I was in a world of fog, shut off from the universe. I did not see any more of the light.*[19]

Slocum was a great showman and storyteller, so it would be unwise to take everything he wrote at face value, but there is no reason to doubt the main elements of his account in *Sailing Alone Around the World*. First he called at Horta in the Azores, and then Gibraltar, before he headed south down the Atlantic and—like so many before him—fought his way through the Straits of Magellan. Having "anchored and weighed many times" and after beating "many days against the current," Slocum at last reached Port Tamar, with Cape Pillar in sight to the west. "Here I felt the throb of the great ocean that lay before me. I knew now that I had put a world behind me, and that I was opening out another world ahead."[20]

Heading out into the Pacific beyond Cape Pillar, however, the *Spray* ran into very heavy weather:

> *There was no turning back even had I wished to do so, for the land was now shut out by the darkness of night. The*

wind freshened, and I took in a third reef. The sea was confused and treacherous. . . . "Everything for an offing," I cried, and to this end carried on all the sail she would bear . . . but on the morning of March 4 the wind shifted to the southwest, then back suddenly to northwest, and blew with terrific force. . . . No ship in the world could have stood up against so violent a gale. . . .[21]

Slocum had no choice but to run before the wind, and so the *Spray* drove southeast, as though to round the Horn:

The first day of the storm gave the Spray her actual test in the worst sea that Cape Horn or its wild regions could afford, and in no part of the world could a rougher sea be found than at this particular point, namely, off Cape Pillar, the grim sentinel of the Horn.[22]

In these conditions Slocum could only trust to DR, and on the fourth day of the gale, when he thought he was nearing Cape Horn, he saw through "a rift in the clouds a high mountain, about seven leagues away on the port beam." He now headed for the land, which appeared as an island in the sea: "So it turned out to be, though not the one I had supposed." The weather was moderating, but the sea conditions were still bad and night closed in before the *Spray* reached the land, leaving Slocum to feel the way in "pitchy darkness." He saw breakers ahead and stood offshore, but was "immediately startled by the tremendous roaring of breakers again ahead and on the lee bow." He was puzzled, and no doubt alarmed, for there should have been no broken water where he supposed himself to be, and he must have realized that his navigation had gone seriously wrong. He stood off and on repeatedly, but each time he closed the coast he was confronted by breakers:

In this way, among dangers, I spent the rest of the night. Hail and sleet in the fierce squalls cut my flesh till the blood trickled over my face; but what of that? It was daylight, and the sloop was in the midst of the Milky Way of the sea, which is northwest of Cape Horn, and it was the white breakers of a huge sea over sunken rocks which had threatened to engulf her through the night. It was Fury Island I had sighted and steered for, and what a panorama was before me now and all around! . . . This was the greatest sea adventure of my life. God knows how my vessel escaped.[23]

Slocum was not exaggerating, as is evident from the description in the 1993 edition of the *South America Pilot* book, which uses the Spanish name for the Milky Way:

Via Láctea . . . so named because it is white with spume . . . consists of innumerable rocks, some above water, on which the sea continually breaks. This area, which includes Rocas Júpiter and Rocas Neptuno, should be avoided as it is little known and extremely dangerous.[24]

The *Spray* at last reached a safe anchorage behind some small islands that sheltered her in smooth water. Then Slocum climbed the mast to survey the wild scene astern:

The great naturalist Darwin looked over this seascape from the deck of the Beagle, and wrote in his journal, "Any landsman seeing the Milky Way would have a nightmare for a week." He might have added, "or seaman" as well.[25]

Slocum was perhaps quoting from memory. Darwin's actual words were even more awestruck: "One sight of such a coast is enough to make a landsman dream for a week about shipwrecks, peril, and death."[26] And we have already heard FitzRoy's professional judgment of "Breaker Bay," which lies close by. Slocum now returned to the relatively sheltered waters of the Straits of Magellan via the Cockburn Channel. If he was chastened by his narrow escape from shipwreck, the effects quickly wore off. Soon he was delighted by his success in defending himself from marauding Tierra del Fuegians. These he claimed to have discouraged by placing upturned carpet tacks across the *Spray*'s decks. Anything but politically correct, Slocum cheerfully boasted:

> *I had no need of a dog; they howled like a pack of hounds. I had hardly use for a gun. They jumped pell-mell, some into their canoes and some into the sea, to cool off, I suppose, and there was a deal of free language over it as they went.*[27]

After many setbacks and delays, Slocum finally escaped the confines of the Straits on April 13, 1896, and headed for the island of Juan Fernández, where he visited the cave of Alexander Selkirk, one of the inspirations for Defoe's *Robinson Crusoe*. He found it "dry and inhabitable," in a "beautiful nook" sheltered by high mountains.[28] From there he sailed for the Marquesas Islands, 3,500 nautical miles to the north and west in the middle of the Pacific. It was some time before he picked up the trade winds, but

> *when they did come they came with a bang, and made up for lost time; and the Spray, under reefs, sometimes one, sometimes two, flew before the gale for a great many days, with a bone in her mouth. . . . My time was all taken up those days—not by standing at the helm; no man, I think, could stand or sit and steer a vessel round the world: I did*

Colin McMullen standing at the stern of *Saecwen*, holding the sextant and preparing to take a sight.

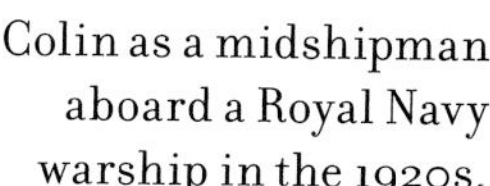

Colin as a midshipman aboard a Royal Navy warship in the 1920s.

Saecwen on the coast of Maine, with the author standing at the bow.

The Nebra Sky Disc, one of the earliest-known representations of the cosmos. Dating from about 1,600 BCE, it was discovered in Germany in 1999. The group of seven stars represents the constellation of the Pleiades as it would have appeared at that date.

Star-trails in the northern sky: a long-exposure photograph taken over the Grand Canyon showing the paths traced by each star. Polaris, in top center, being almost stationary, appears as a dot, while the stars at the greatest angular distance from it leave the longest trails.

The mutineers turning William Bligh and his crew from the *Bounty*. April 29, 1789. A cutlass is being thrown to the castaways from the great cabin, and breadfruit plants are growing in the pots hung over the stern. Aquatint by Robert Dodd, c. 1790.

A pastel portrait of William Bligh in 1791, by John Russell RA.

Nevil Maskelyne, at about forty-four. A chalk drawing attributed to John Russell RA.

One of the very first sextants, made in London by John Bird about 1758. With a radius of 18¼ inches, it is large in comparison with modern instruments. It is also unusual in having a pole that fits into a socket on the observer's belt to help support its weight.

21

H	K	F	Cours.	Winds	Sound.s	☿ 15
1	4		W.t	ESE		moderate breezes and fair – Saw moderately high land WSW
2	3	7				Saw more land making like is.
3	3	5				Saw a reef of rocks between us and the shore
4	3	4				
5	3	5			no gr. at 140	found the reef extended N.o & S.o as far as we could perceive – hauld off making all sail and kept a good look out
6	3	5	NNE	E.t		all night to leew.d being doubtful of our weathering the rocks
7	3		NbE	EbN	– – 140	
8	3					
9	3	2				
10	3	3				
11	3					
12	3					tackd
13	2		SSE	E.t		
14	Calm		Ships head SSE			
15			SE			

At 4 a.m. heard and saw breakers upon the lee bow close too, lowerd the yawl, sent her ahead to tow off mean time empd getting the longboat out; the ship all this time nearing the rocks fast by means of a flood tide and SE swell – The Pinnace having sufferd much in service on this coast was now under repair, however the Carpenters fastend an upper strake on and she with all the other boats was ahead towing at 3/4 past 5 at which time the ship was within 40 yards of the breakers and no gr. at 120 fath. – Notwithstanding the ship might be said to be within the swell of the surf the boats and a light air run her off 1/2 a cables length from that impending danger – soon after saw an opening in the reef, sent a Mate to examine it – A signal being made from the boat for Anchorage & no hopes of clearing the reef steerd or rather towd for the opening but meeting with a strong ebb tide and the Mate returning without a satisfactory account towd off again and by the help of the ebb got clear of another point of the reef farther N.o – same time saw a point of a reef NEbN – At Noon Calm and Clear low water – the reef dist.t 1/2 a mile and small hopes of getting Clear.

Lat. Ob. 12..37' S.o

Watch	Ꞓ ☉'s L.L.F	Forwood ☉'s [illegible] L.F	Ꞓ ☽ & ☉
21 45.9	46 24	48 56	54 54
46.35	42	34	53 45
48.0	47 0	19	53 30
46.37	46 42	36 20	53 45
49.45	47 20	6	53 0
51.51	37	47 56	52.45
53.3	56	46	52.15
51 33	47 37 40	56 0	52.40
56 49	48 19	28	51 45
57 44	39	15	51 45
58 18	56	0	51.15
57 37	48.38	47.14.20	51.35

Rem.s These Obs.s very good the Limbs very distinct, a good Horizon We were about a 100 Yards from a Reef where we expected the Ship to strike every minute it being Calm & no soundings the swell heaving us right on.

		Calm / ESE / E.t / EbN	♃ 16

Calms and light airs with Clear weat.r – an opening being discovered in a bend of the reef Lieut. Hickes went to examine it – at 2 he returnd with a favourable acco.t of it – Nothing but danger appearing on all sides it was resolved to attempt this passage to secure the ship till there was wind and opportunity to command her – accordingly towd (with the assistance of the Sweeps out of the Gunroom ports which had been working from 6 o'clock) short round – steerd WbS 2 m.s to the mouth of the opening and SWbW ½ W 2 m.s thro' the opening having a rapid flood setting us in – Sound.s varble from 13 to 30 fath. foul ground ~ ½ past 3 had a steady light breeze at E.t (all this time the Boats ahead, 2 towing and 2 sounding) – ½ past 4 came too with the best Bow.r in 19 fath. Old Coral & Shells veerd to 1/3 of a Cable, the opening NE ½ E 2 m.s – found the flood here setting strong from ESE. ~ A.M. Carpenters empd on the Pinnace – sent the rest of the boats to the reef to get shell fish or turtle if possible.

Variation Ampl. and azym.s . . . . 4.o 09' E.t

Latit Obs.d . . . . 12.. 38 S.o

♀ 17 moderate breezes and Clear pleasant weat.r – at ease (low water) lookd well out for shoals ~ at 4 the boats returnd with 270 lbs of fish – People variously employd.

H	K	F	Cours.	Winds		
17						began to heave up
18						weighd and came to sail ~ The Yawl ahead
19	2	6	NWbW	SE	16.14.14.12.18	an opening in the land bore S44 W 5½ leag. –
20	3				13.11.12.15	
21	2	6		ESE	17.18.20	Saw an island and 2 rocks NWbN
22			NNW / NW½W		24.26.27	
23	2	2			23.20.17.15 / 16.15.22.27	Moderate breezes and smooth water – Outer is.d ahead NW 3 or 4 leag. – a sandy is.d on a small reef ESE ½ E – Outer or No.m.t land of the main N.73°W 5 or 6 leag.
24	2		NW½N		22.19.17.16 / 17.19.22.27	Saw a reef in shore SW – yawl ahead sounding

Lat Obs: 12.. 28' S.o

A page from a journal showing Charles Green's lunar distance observations taken aboard the *Endeavour* off the Great Barrier Reef (*box, center left*). The latitude is prominently recorded (*center right*) and beneath it Green notes the unusual circumstances: "These Obs[ervations] very good the Limbs very distinct, a good Horizon. We were about a 100 yards from a Reef where we expected the Ship to strike every minute it being Calm & no soundings the swell heaving us right on."

Louis-Antoine de Bougainville. Lithograph after a portrait by Zépherin Belliard, date unknown.

Jean-François de Galaup, comte de La Pérouse. This fine portrait by Jean-Baptiste Greuze—probably painted in the 1780s—skillfully suggests the sitter's combination of determination and charm.

A portrait, believed to be of George Vancouver, by an unknown artist, 1796–98.

The Caneing in Conduit Street. — Dedicated to the Flag Officers of the British Navy.

"The Caneing in Conduit Street," by James Gillray. Vancouver shouts "Murder!" and calls for his brother (in the center) to defend him, while Pitt (Lord Camelford) says, "Give me Satisfaction, Rascal!"

James Cook by John Webber. The official artist on Cook's third voyage, Webber knew Cook better than any other artist and had seen him in his element. This touching portrait—painted in 1781, after Cook's death—shows us a more vulnerable, less heroic figure than some other famous images of the great mariner.

Matthew Flinders, while held in captivity by the French authorities on Mauritius, painted by the amateur artist Antoine Toussaint de Chazal de Chamerel. Though unsophisticated, this portrait clearly reveals Flinders's stubborn and prickly character.

Robert FitzRoy. Albumen print by unknown photographer, 1860s.

Alcyone, named after the brightest star in the constellation of the Pleiades, off the coast of Brittany.

A typical marine chronometer, similar to the one we used aboard *Saecwen*; the smaller dial at the top indicates when the mechanism needs to be rewound.

Joshua Slocum, by an unknown photographer (undated).

Slocum's yawl, the *Spray*, in Sydney Harbor, 1896. Unknown photographer.

Frank Worsley aboard the *Endurance*. Photograph by Frank Hurley.

The *Endurance* crushed by the Antarctic ice and sinking as sledge dogs look on. Photograph by Frank Hurley.

better than that; for I sat and read my books, mended my clothes, or cooked my meals and ate them in peace. I had already found that it was not good to be alone, and so I made companionship with what there was around me, sometimes with the universe and sometimes with my own insignificant self; but my books were always my friends, let fail all else. . . .[29]

Slocum generally says little about his navigational methods, and it sounds as if he relied on DR (he admits to using a "rotator log"—just as we did in *Saecwen*) and latitude sailing for much of the time, but on this occasion it was his knowledge of lunars that enabled him to make a safe landfall. Day after day he sailed with a following wind, marking the position of the *Spray* on the chart more by intuition than "slavish calculations." For a whole month his vessel held her course true without so much as a light in the binnacle. Every night he saw the Southern Cross shining out abeam. Every morning the sun rose astern and every evening it went down ahead of him. He wished for no other compass to guide him, as "these were true," but at last he reached for his sextant:

To cross the Pacific Ocean, even under the most favorable circumstances, brings you for many days close to nature, and you realize the vastness of the sea. Slowly but surely the mark of my little ship's course on the track-chart reached out on the ocean and across it, while at her utmost speed she marked with her keel still slowly the sea that carried her. On the forty-third day from land,—a long time to be at sea alone,—the sky being beautifully clear and the moon being "in distance" with the sun, I threw up my sextant for sights. I found from the result of three observations, after long wrestling with lunar tables, that her

Fig 9: Admiralty chart of Strait of Magellan, first published in the 1830s and based on surveys by King, FitzRoy, and Stokes.

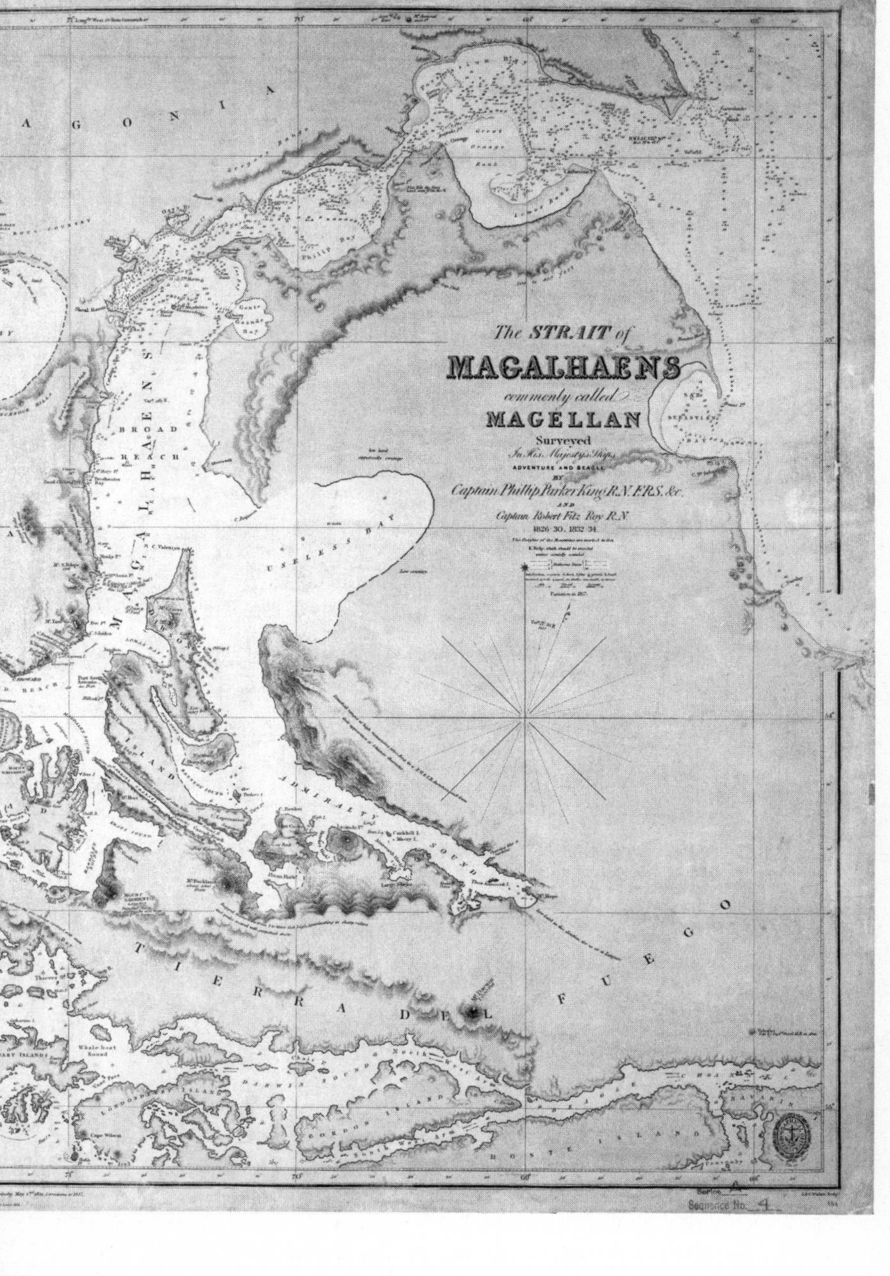
The STRAIT of
MAGALHAENS
commonly called
MAGELLAN
Surveyed
In His Majesty's Ships
ADVENTURE AND BEAGLE
BY
Captain Phillip Parker King R.N. F.R.S. &c.
AND
Captain Robert Fitz Roy R.N.
1826-30. 1832-34.
A G O N I A
BROAD REACH
M A G A L H A E N S
USELESS BAY
ADMIRALTY SOUND
T I E R R A D E L F U E G O
DARWIN SOUND
GORDON ISLAND
HOSTE ISLAND
Whale-boat Sound
Great Orange Bank
Philip Bay
Series A
Sequence No. 4

longitude by observation agreed within five miles of that by dead-reckoning.

This was wonderful; both, however, might be in error, but somehow I felt confident that both were nearly true, and that in a few hours more I should see land; and so it happened, for then I made the island of Nukahiva, the southernmost of the Marquesas group, clear-cut and lofty. The verified longitude when abreast was somewhere between the two reckonings; this was extraordinary.*[30]

Remarkable indeed, and very different from the result Slocum had obtained when he first performed the calculations. The first set of sights had put the *Spray* hundreds of miles west of her DR position, and Slocum knew that this could not be correct. So he took another set of observations with the utmost care, but the average result was about the same as that of the first set. He asked himself why, with his "boasted self-dependence," he had not done better and, never lacking self-confidence, decided to look for a discrepancy in the tables. There he found that an important logarithm was in error:

It was a matter I could prove beyond a doubt, and it made the difference as already stated. The tables being corrected, I sailed on with self-reliance unshaken, and with my tin clock fast asleep. The result of these observations naturally tickled my vanity, for I knew that it was something to stand on a great ship's deck and with two assistants take lunar observations approximately near the truth. As one of the poorest of American sailors, I was proud of the little

---

* This is a mistake: Nuku Hiva is actually one of the more northerly members of the Marquesas Islands. Perhaps Slocum meant to write Hiva Oa.

> *achievement alone on the sloop, even by chance though it may have been.*[31]

Slocum described how he was now in harmony with the world around him, as if carried on "a vast stream" where he "felt the buoyancy of His hand who made all the worlds":

> *I realized the mathematical truth of their motions, so well known that astronomers compile tables of their positions through the years and the days, and the minutes of a day, with such precision that one coming along even five years later may, by their aid, find the standard time of any given meridian on the earth. . . . The work of the lunarian, though seldom practiced in these days of chronometers, is beautifully edifying, and there is nothing in the realm of navigation that lifts one's heart up more in adoration.*[32]

Errors in the tables on which navigators relied were rare, but they were not unknown, especially in the days when they depended on human calculators. In the late eighteenth century, the great Nathaniel Bowditch (1773–1838)—whose name is still attached to the standard American navigation manual—discovered numerous mistakes in John Hamilton Moore's *Practical Navigator*. Although most were unimportant, the incorrect listing of the year 1800 as a leap year meant that the solar declination for March 1, 1800, was out by 22 minutes—an error that led to at least one shipwreck.[33] Nevertheless, it says much not only for Slocum's self-confidence, but also for his mathematical skills, that he was able to find the faulty logarithm in the tables he was using and then rectify it. A less experienced navigator might well have assumed that the mistake was his own and given up. It is also interesting that Slocum acknowledges the possibility that the accuracy of his lunar-based fix may just have been a fluke.

Slocum crossed the Pacific by way of Samoa—where he visited the house of his hero Robert Louis Stevenson, who had only recently died there—and when he reached Australia discovered that he was "news." He was now able to earn money by giving lectures to audiences keen to hear about his adventures. Slocum then headed north from Sydney and passed through the Torres Strait before calling at the Cocos Islands and Mauritius en route to South Africa. By the time he returned home in June 1898 he was a celebrity, and his account of the voyage (published in 1900) has since become a classic. His latter years were, however, darkened by a charge of rape brought against him in 1906 by a girl of twelve. He was let off with a reproof from the judge for his "great indiscretion" and banned from ever again visiting the town where the offense had occurred.[34] Slocum disappeared at sea after setting sail from Martha's Vineyard on a single-handed voyage to the Amazon in November 1908.

Chapter 16

Endurance

Day 18: Fog and drizzle descend on us. But still doing 6½ knots until mid-morning. Slept for an hour or so and when I got up there was fog with visibility down to half a mile or less. Wind eased slightly slowing us to 5½ knots but even so, by noon we had covered 155 miles by the log. Our best day's run. No sights but noon position by DR: 47°27' N, 20°21' W. Land's End only 600 miles away.

Talked with Colin about naval warfare and how Captain Cook conducted his surveys. The sextant was useful not just for celestial work but also for fixing positions by horizontal angles—triangulation.

Saw another Portuguese man-of-war. A butterfly—a Large White I think—flapped by strongly. Where was it going? Where was it from?

Played Salvoes with Alexa. Fog persisted in evening and we all felt tired and bored. Wind almost died and we trickled along at 3–4 knots. Good stew for supper with whisky and rice pudding.

Day 19: A better breeze and making 4–5 knots on 090° under a NW wind force 2–3. The fog began to lift at 0730 and we managed to take some sights which I worked out and plotted. The position line strongly suggested that we'd gone south of our course by perhaps 5–10°. Later in the day we took*

---

* Undetected, this error would have meant making our landfall on the French coast somewhere on the southwest coast of Brittany.

the spare compass up forward and confirmed that the steering compass was deviating by this amount to the south.

A very calm day. Some sun finally broke through and I put on my shorts again. Then the wind dropped almost completely. A strong swell was running and, with the sails flogging uselessly, we rolled miserably for an hour or two before reluctantly starting up the motor. Once we were moving again, it was more comfortable as we no longer had to brace ourselves constantly to avoid being thrown around. Reading Erskine Childers' Riddle of the Sands. *Very jingoistic but good descriptions of small boat sailing. By now getting BBC radio loud and clear all day and heard news of Nixon's resignation. Got time signal—chronometer behaving well though actually not much more accurate than my wristwatch. A very hot day at home it seems.*

Noon position 47°19' N, 17°23' W and 119 miles run. An enjoyable quiet sail but a test for our patience with England only 500 miles off now.

The Atlantic gale we experienced in *Saecwen* was nothing compared to the storms that many sailors have had to endure—especially in the Southern Ocean, which encircles Antarctica. Here strong, often very strong, westerly winds blow almost without intermission throughout the year across thousands of miles of ocean, and enormous seas have time and space to build up. This was the scene in 1916 of an extraordinary epic of survival that illustrates the vital importance both of good seamanship and of the skill that comes from years of practice with a sextant.

The hero of this story was Frank Worsley, a merchant seaman from New Zealand, who had, like Slocum before him, learned his trade under sail. In 1888, at the age of sixteen, Worsley joined a clipper ship sailing to England, a passage that gave him his first experience of a Southern Ocean gale. He described the "terrific grandeur of the scene and the danger of toppling seas rolling on, ridge behind ridge" as the ship headed toward the Diego Ramírez Islands:

It is too late to heave-to; the ship might founder in the attempt. It is necessary to drive her to prevent her from pooping or broaching-to, and so the captain cracks on with four full sails and reefed mainsail and defying the gale, refuses to lower the topsail.[1]

Worsley rose steadily up the ladder and was given his first command in 1901, a trading schooner in which he travelled widely in the South Pacific. His experience of handling small boats among the islands was to prove invaluable later. He joined the Royal Naval Reserve and in due course settled in London, where, in 1914, he met Sir Ernest Shackleton. "The Boss," as he was known to his crew, was planning a grand expedition to cross the Antarctic continent, and he was looking for someone to command the ship that was to take him there: the *Endurance*. His first choice had, perhaps presciently, turned the job down and Worsley was appointed almost by chance:

One night I dreamed that Burlington Street was full of ice blocks, and that I was navigating a ship along it—an absurd dream. Sailors are superstitious, and when I woke up next morning I hurried like mad into my togs, and down Burlington Street I went. I dare say that it was only a coincidence, but as I walked along, reflecting that my dream had certainly been meaningless and uncomfortable and that it had cost me time that could have been used to better purpose, a sign on a door-post caught my eye. It bore the words "Imperial Trans-Antarctic Expedition," and no sooner did I see it than I turned into the building with the conviction that it had some special significance for me.[2]

Worsley and Shackleton met briefly. "The moment I set eyes on him," Worsley later recalled, "I knew that he was a man with whom I should be proud to work," and Shackleton must have taken a liking to Worsley because he hired him on the spot. In August 1914, just as World War I was breaking out, Worsley was ready to set sail. As a reservist, he, like many other members of the crew (twenty-eight men in all), expected to be called up, and Shackleton in fact put the ship at the disposal of the Royal Navy. However, word came down from the First Lord of the Admiralty, Winston Churchill, that the expedition should go ahead. Everyone believed that the war would soon be over.

The *Endurance* reached Buenos Aires in October, where Shackleton joined her, and then called at South Georgia. Although there were alarming reports of the extent of the ice pack, the *Endurance* headed south on December 5 and was soon pounding her way through the ice floes.[3] On January 19, 1915, their slow, zigzag progress finally ceased when the ship was only sixty miles from her planned destination on the Antarctic coast. Sextant observations placed her in 76°34' South, 31°30' West, and land was faintly visible in the east,[4] but despite heroic efforts to free her, the *Endurance* was inextricably beset in pack ice. On January 22 they reached the southernmost point of their drift and the southern summer was over. The temperature was dropping, the sea was freezing solidly around them, and Shackleton "could not now doubt that the *Endurance* was confined for the winter":

> *We must wait for the spring, which may bring us better fortune. . . . My chief anxiety is the drift. Where will the vagrant winds and currents carry the ship during the long winter months that are ahead of us?*[5]

On February 24 they ceased to observe ship routine and the *Endurance* became a winter station, with the many sledge dogs kennelled in igloos on the ice. Seals provided fresh meat for dogs and men alike. The accommodation aboard the ship was adapted so that the crew could survive the winter in reasonable comfort. Although they kept themselves busy exercising the dogs, making scientific observations, and organizing games, the long months of drift must have been extraordinarily trying. On May 1, 1915, they said good-bye to the sun and Shackleton wrote in his journal:

> *One feels our helplessness as the long winter night closes upon us. By this time, if fortune had smiled upon the Expedition, we would have been comfortably and securely established in a shore base. . . . Where will we make a landing now? . . . Time alone will tell.*[6]

Their position on May 2 was 75°23' South, 42°14' West—about 170 nautical miles from the nearest part of the Antarctic coast. An "Antarctic Derby" sledge race was held on June 15 and Midsummer's Day (midwinter in the Antarctic) was celebrated with a big party.[7] By now the ship was drifting steadily northward with the ice pack. A prolonged, severe storm in July led to the break-up of the ice around the ship on August 1.[8] Floes forced their way beneath her keel and others crashed into her alarmingly. Shackleton readied the boats in case the crew needed quickly to abandon ship, but the *Endurance* survived this assault. A sextant sight of the star Canopus on August 3 confirmed their northward progress. Wonderful mirages appeared over the ice, as Shackleton recorded in his journal:

> *The distant pack is thrown up into towering barrier-like cliffs, which are reflected in blue lakes and lanes of water at their base. Great white and golden cities of Oriental appearance at close intervals along these cliff tops indicate distant bergs, some not previously known to us. Floating*

> *above these are wavering violet and creamy lines of still more remote bergs and pack. The lines rise and fall, tremble, dissipate, and reappear in an endless transformation scene. The southern pack and bergs, catching the sun's rays, are golden, but to the north the ice masses are purple. Here the bergs assume changing forms, first a castle, then a balloon just clear of the horizon, that changes swiftly into an immense mushroom, a mosque, or a cathedral.*[9]

In September the ship was again threatened by the surrounding ice, and at the end of the month, as "the roar of the pressure grew louder," Shackleton could see an area of disturbance in the ice pack that was rapidly approaching:

> *Stupendous forces were at work and the fields of firm ice around the Endurance were being diminished steadily. . . . The ship sustained terrific pressure on the port side forward. . . . It was the worst squeeze we had experienced. The decks shuddered and jumped, beams arched, and stanchions buckled and shook. I ordered all hands to stand by in readiness for whatever emergency might arise.*[10]

Worsley warmly praised the *Endurance*, whose behavior in the ice had been "magnificent":

> *She has been nipped with a million-ton pressure and risen nobly, falling clear of the water out on the ice. She has been thrown to and fro like a shuttlecock a dozen times. She has been strained, her beams arched upwards, by the fearful pressure; her very sides opened and closed again as she was actually bent and curved along her length, groaning like a living thing. It will be sad if such a brave little craft*

> *should be finally crushed in the remorseless, slowly strangling grip of the Weddell pack after ten months of the bravest and most gallant fight ever put up by a ship.*[11]

The ship stood firm, and warmer weather gave them hope that they might yet escape the grip of the ice. During October they survived further contests with the ice, but at the end of that month they were pinched between three different pressure ridges, and at last the *Endurance* began to leak heavily. Worsley had the vital task of clearing one of the pumps, frozen solid deep in the bilges:

> *This is not a pleasant job. . . . We have to dig a hole down through the coal while the beams and timbers groan and crack all around us like pistol shots. The darkness is almost complete, and we mess about in the wet with half-frozen hands and try to prevent the coal from slipping back into the bilges. The men on deck pour buckets of boiling water from the galley down the pipe as we prod and hammer from below, and at last we get the pump clear, cover up the bilges to keep the coal out, and rush out on deck, very thankful to find ourselves safe again in the open air.*[12]

For the moment they could keep the leaks under control, but the end was approaching. On October 26 something very strange occurred, which to men in their situation must have seemed ominous. Eight emperor penguins suddenly appeared out of a crack in the ice, just as the pressures on the ship were reaching their destructive climax:

> *They walked a little way towards us, halted, and after a few ordinary calls proceeded to utter weird cries that sounded like a dirge for the ship. None of us had ever before*

heard the emperors utter any other than the most simple calls or cries, and the effect of this concerted effort was almost startling.[13]

The next day, Shackleton was finally forced to abandon ship. He found it hard to write what he felt:

To a sailor his ship is more than a floating home, and in Endurance I had centred ambitions, hopes and desires. Now, straining and groaning, her timbers cracking and her wounds gaping, she is slowly giving up her sentient life at the very outset of her career. She is crushed and abandoned after drifting more than 570 miles in a northwesterly direction during the 281 days since she became locked in the ice.[14]

Now they could only watch helplessly as the *Endurance* was at last destroyed:

The twisting, grinding floes were working their will at last on the ship. It was a sickening sensation to feel the decks breaking up under one's feet, the great beams bending and then snapping with a noise like heavy gunfire. . . . The plans for abandoning the ship in case of emergency had been made well in advance, and men and dogs descended to the floe and made their way to the comparative safety of an unbroken portion of the floe without a hitch. Just before leaving, I looked down the engine room skylight as I stood on the quivering deck, and saw the engines dropping sideways as the stays and bed plates gave way. I cannot describe the impression of relentless destruction that was forced upon me as I looked down and around. The floes,

> *with the force of millions of tons of moving ice behind them, were simply annihilating the ship.*[15]

They were in a tight spot, and it was not clear what they should do. Shackleton was in favor of dragging the ships' boats over the ice to the nearest land, now some 350 miles distant, where he hoped to find a store of emergency supplies that he had left there some years earlier, but this preposterous plan soon proved impracticable. Morale was understandably low, and there was almost a mutiny, but Shackleton managed to maintain discipline. They now had no choice but to follow Worsley's more sensible advice. This was to allow the ice to carry the whole party and all the equipment they had saved from the wreck as far north as possible and then take to the boats. Accordingly they settled down to endure five months encamped on the ice pack.

On April 9, 1916, at the beginning of the Antarctic autumn, after drifting some six hundred miles northward, Shackleton and his men—twenty-eight in all—launched the three surviving boats as the floes began to break up around them. Sextant sights now indicated that they were about sixty miles southeast of Elephant Island, itself about 480 miles southeast of Cape Horn.[16]

Lanes of open water began to appear on the western horizon, so they packed everything ready for launching and struck the tents. The floe on which they were encamped suddenly cracked, right across the camp and through the site of Shackleton's tent, which he had only just vacated. Having launched the boats and pulled clear into an area of partially open water, they were nearly caught by a heavy rush of wind-driven pack ice that drove toward them at an alarming pace:

> *Two dangerous walls were converging as well as overtaking us, with a wave of foaming water in front. We only just*

> *managed by pulling our damnedest for an hour to save ourselves and the boats from being nipped and crushed. It was a hot hour in spite of the freezing temperature.*[17]

That night, while camped out on a floe-berg with the boats pulled out beside them, the heavy swell cracked the ice beneath one of the tents, ripping it in two. One of its occupants fell into the water, in his sleeping bag, but in spite of the darkness Shackleton managed with one heave to pull the man and his bag up onto the ice. A moment later "the halves of the floe swung together in the hollow of the swell with a thousand-ton blow."[18]

By April 11, the swell had reached a "tremendous height," and Worsley described the "magnificent and beautiful" but at the same time fearsome spectacle:

> *Great rolling hills of jostling ice sweeping past us in half-mile-long waves. A few dark lines and cracks sharply contrasted against the white pack were the only signs of the sea beneath. But it was a sight we did not like, for the floes were thudding against our floe-berg with increasing violence. Our temporary home was being swept away at an unpleasantly rapid rate.*[19]

Luckily they drifted into a patch of open water and launched the boats again in a hurry. A "cold, wet, rotten night—all hands wet and shivering" followed, as they dodged through the ice pack in temperatures that fell to 25 degrees Fahrenheit below freezing (about −14 Celsius). At dawn on April 12 they set sail again, and "as soon as the horizon cleared" Worsley took sextant sights for longitude, followed by a mer alt at noon for latitude. But when he worked out their position, it was a "terrible disappointment." He had previously told Shackleton that they had made thirty miles' progress toward Elephant Island, but the sights proved that they were, in fact, thirty miles farther away from their destination:

> *Though the responsibility must rest on the leader of an expedition, I can never forget my acute anxiety for the next two days. If there was a mistake in my sights, which were taken under very difficult conditions, twenty-eight men would have sailed out into death. Fortunately the sights proved correct. We made the Island ahead fifty hours later.*[20]

After two days of dead reckoning while they worked their way through the ice, and two cold nights spent drifting to reduce the risk of a potentially disastrous collision with a berg, "the peaks and ice uplands of Elephant Island showed cold and gloomy, thirty-five miles to the NNW."[21] They were exactly on the bearing Worsley had predicted and Shackleton praised him warmly, but he modestly admitted that there had been "a large element of luck in making this good landfall." As they approached the island on April 15 they were separated from the two other boats and found themselves in a dangerous tidal race that nearly sank them as the "seas leapt simultaneously over bows, beam and stern." By bailing desperately they just managed to stay afloat, but many of the crew were suffering from frostbite, and thirst too was a big problem as they had run out of fresh water.[22]

Worsley had been steering for twenty straight hours and "constant peering to windward to ease her into the heaviest seas, and the continuous dash of salt-water into my face had almost bunged up" his eyes. He handed over the tiller and fell so fast asleep that his crewmates thought he was dead. When they began closing the land, they could wake him only by giving him a sharp kick in the head.

The prospect was imposing: "Great crags thrust up through the snow. A wall of ice faced us with high seas rolling along its base."[23] Now they were running before a gale, with heavy seas coming up astern that threatened to fill the boat, but they reached safety in the lee of the island and, as they approached a rocky beach, discovered to their delight that their companions had arrived before them. They had been at sea for one hundred hours—four full days and nights—and during this time Worsley had slept for only one hour. Shackleton had not slept at all.[24] Even now they were far from safe. The beach proved untenable, and they had to move along the coast—a risky maneuver in the course of which strong offshore winds very nearly carried them out to sea. At last, after rowing desperately to save their lives, they reached the beach that was to be their new home "in the flaming glow of an amazingly beautiful but stormy sunset." Their tents had been torn to shreds, so they now took shelter under the three upturned boats.[25]

Chapter 17

"These Are Men"

Day 20: Trickled along at 2 to 3 knots all through the night and a heavy-headed awakening at 0400. Grey skies but no fog and by midday a good southerly breeze which set us off at 5 knots or more. Finally on the Great Britain chart. No sun sight but roughly 47°50' N, 15°W at noon. A very exhilarating sail during the afternoon over smooth seas leaving a hissing, pulsing wake behind us. Put clocks forward another hour to match GMT—still an hour behind BST. Tired but happy.

Day 21: Another grey cloud-shrouded dawn—keeping to 6 knots in slight seas. Force 3 from SSE. Passed various ships—they are now becoming more plentiful.

Position at noon 48°46' N, 12° 07' W—I worked out the plot by myself. In the evening the wind dropped off and we passed some fishing boats—a sure sign that land is near. Close now to the edge of the Continental Shelf but still in water a mile deep.

On Elephant Island the southern winter was again approaching, and if the crew were to stand any chance of surviving someone would have to rescue them soon. Shackleton consulted Worsley and they decided that the best option was to try to reach one of the whaling stations on the north coast of the inhospitable island of South Georgia, discovered by Cook on his second voyage. This would entail a passage of eight hundred nautical miles, under sail, across the Southern Ocean in a ship's boat that was only 23.5 feet long—the *James Caird*.* She was equipped with sails, four oars, a Primus stove, a compass, two water-breakers, some red flares, matches, and a small medicine chest, as well as—crucially—the sextant, chronometer, nautical almanac and charts. They extended her deck as best they could with the help of sledge-runners, canvas, and box lids, creating a cramped, low, dark, smelly cabin, where they laid their reindeer-fur sleeping bags (soon to be soaked through) on top of the boxes of stores. The mast from one of the other boats was then bolted to the *James Caird*'s keel to reduce the risk of her breaking her back in "extra heavy seas." Even the expedition artist's oil paints were put to use in making the boat's seams more watertight, being topped off with seal's blood. The crew had no sea boots or oilskins, as these had all worn out or been cut up and reused for other purposes.[1]

Before setting off, Worsley had one absolutely vital chore to perform with his sextant:

> *Every day I watched closely for the sun or stars to appear, to correct my chronometer, on the accuracy of which our lives and the success of the journey would depend. . . . Never a gleam of sun or stars showed through the dull grey or else storm-driven pall of clouds that in these latitudes seems ceaselessly and miserably to shroud the bright blue sky and the cheerful light of sun or moon.*[2]

Shackleton chose to take Worsley with him, "for I had a very high opinion of his accuracy and quickness as a navigator, and especially in the snapping and working out of positions in diffi-

---

* The *James Caird*—named after one of the principal sponsors of the expedition—is preserved at Dulwich College in South London.

cult circumstances—an opinion that was only enhanced during the actual journey."[3] Four other crew members were needed, so he asked for volunteers and selected those he judged best able to withstand the rigors of the impending voyage. At last Shackleton decided that they could wait no longer, and on April 24 they began loading the boat. Luckily the sun came out at the last minute, enabling Worsley to get the crucial sight. Only one chronometer had survived "in good going order" out of the twenty-four with which they had first set out in the *Endurance*.[4]

After almost capsizing in the surf, the *James Caird* set sail on a voyage that was to last sixteen tumultuous days. Shackleton recorded the scene:

> *The men who were staying behind made a pathetic little group on the beach, with the grim heights of the island behind them and the sea seething at their feet, but they waved to us and gave three hearty cheers.*[5]

The first night out, once clear of the ice floes, while the four other members of the crew rested beneath the makeshift deck, Worsley and Shackleton stood watch together:

> *I steered; he [Shackleton] sat beside me. We snuggled close together for warmth, for by midnight the sea was rising, and every other wave that hit her came over, wetting us through and through. Cold and clear, with the Southern Cross high overhead, we held her north by the stars, that swept in glittering procession over the Atlantic towards the Pacific. While I steered, his arm thrown over my shoulder, we discussed plans and yarned in low tones. We smoked all night—he rolled cigarettes for us both, a job at which I was unhandy. I often recall with proud affection memories of those hours with a great soul.*[6]

They survived ten days of gales and had to pump every few hours to keep the small boat afloat. There was water everywhere, and soon almost everything was soaked through. As the air temperature was often well below freezing, they suffered from frostbite, and at one stage they had to clear a dangerous buildup of ice from the deck to prevent the boat capsizing[7]—a terrifying task with nothing to hold on to and no hope for anyone who slipped overboard. Occasionally at night "the great grey shroud" of cloud was torn aside and they checked their course by their "old friend" the star Antares—the red eye in the constellation Scorpio. They had no spare candles to light the compass, and during the whole voyage Worsley was able to obtain sun sights on only four occasions. And it was far from easy to obtain a fix when the opportunities occurred. His navigating books and log were "soaked through, stuck together, illegible and almost impossible to write in" and DR had "become a merry jest of guesswork." Once or perhaps twice a week "the sun smiled a sudden wintry flicker, through storm-torn clouds" and, "if ready for it, and smart," Worsley would catch it:

> *I peered out from our burrow—precious sextant cuddled under my chest to prevent the seas falling on it. Sir Ernest stood by under the canvas with chronometer, pencil and book. I shouted "Stand by" and knelt on the thwart—two men holding me up on either side. I brought the sun down to where the horizon ought to be and as the boat leapt frantically upward on the crest of a wave snapped a good guess at the altitude and yelled, "Stop." Sir Ernest took the time, and I worked out the result. Then the fun started! Our fingers were so cold that he had to interpret his wobbly figures—my own so illegible that I had to recognize them by feats of memory. . . . My navigation books had to be half-opened, page by page, till the right one was reached, then*

opened carefully to prevent utter destruction. The epitome [*containing all the necessary tables for the calculations*] *had had the cover, front and back pages washed away, while the Nautical Almanac shed its pages so rapidly before the onslaught of the seas that it was a race whether or not the month of May would last to South Georgia. It just did, but April had vanished completely.*

The huge swells of the Southern Ocean often threatened to engulf them:

the highest, broadest and longest in the world, they race in their encircling course until they reach their birthplace again, and so reinforcing themselves sweep forward in fierce and haughty majesty, four hundred, a thousand yards, a mile apart in fine weather, silent and stately they pass along. Rising forty or fifty feet or more from crest to hollow they rage in apparent disorder during heavy gales.[8]

One night when Shackleton was on watch, Worsley heard him shout, "It's clearing, boys!" and then immediately: "For God's sake, hold on! It's got us!"

The line of white along the southern horizon that he had taken for the sky clearing was, in fact, the crest of an enormous sea. I was crawling out of my bag as the sea struck us. There was a roaring of water around and above us—it was almost as though we had foundered. The boat seemed full of water. We . . . seized any receptacle we could find and pushed, scooped and baled [sic] *the water out for dear life. While Shackleton held her up to the wind, we worked like madmen, but for five minutes it was uncertain whether we would succeed or not . . . With the aid of the*

> *little home-made pump and two dippers it took us nearly an hour to get rid of the water and restore the boat to her normal state of only having a few gallons of water washing about the bilges through the stones and shingle.*[9]

On the tenth day, the gale moderated and "Old Jamaica"—the sun—showed his face. Worsley's few sights showed they had run 444 miles from Elephant Island—more than halfway to South Georgia. The next day was fine, with some blue sky and a moderate sea, and further sights were possible. Worsley had now found a better way of taking a sight, by sitting on the deck with one foot jammed between the mast and the halyards and the other braced against the shrouds that supported the mast. His fix put them 496 miles from Elephant Island.[10] On the fourteenth day, after being hove-to for twelve hours in yet another gale, Worsley was more anxious than ever to get a sight as he reckoned they must be nearing their destination:

> *At 9.45 a.m. the sun's limb was clear, but it was so misty that I kept low in the boat to bring the horizon closer, and so a little clearer. . . . At noon the sun's limb was blurred by a thick haze, so I observed the centre for latitude. Error in latitude throws the longitude out, more so when the latter is observed, as now, too near to noon. I told Sir Ernest that I could not be sure of our position to ten miles, so he would not agree to my trying to weather the NW end of South Georgia, for fear of missing it. We then steered a little more easterly, to make a landfall on the west coast.*[11]

Just before dark, when they were still eighty miles offshore, they joyfully spotted a piece of kelp. At dawn on the fifteenth day, they saw more pieces of seaweed, and Worsley looked anxiously for the sun. He was conscious that his navigation had been, "perforce, so extraordinarily crude that a good landfall could hardly be looked for." Later they spotted a shag—a clear sign that land was near, since these birds rarely travel far from the coast. Then, just after noon, the fog cleared and they sighted land—"a towering black crag, with a lacework of snow around its flanks. One glimpse and it was hidden again."[12] Worsley had somehow managed a perfect landfall.

Their troubles, however, were far from over. Not only were the charts of this uninhabited side of the island incomplete, but the *James Caird* was now caught in a terrific storm. She remained hove-to till 2 P.M., when through "a sudden rift in the storm-driven clouds" her crew saw "two high, jagged crags and a line of precipitous cliffs and glacier fronts" astern of them. They were being blown onshore, in the most dangerous and unknown part of the coast—the stretch between King Haakon Sound and Annenkov Island.* The nearest settlements were the whaling stations on the north coast of South Georgia—on the far side of a range of peaks and glaciers. As the small boat drove inshore it seemed that "only three or four of the giant deep-sea swells" separated them from "the cliffs of destruction—the coast of death." If they could but have appreciated it, "a magnificent, awe-inspiring scene" lay before them.

> *The sky all torn, flying scud—the sea to wind'ard like surf on a shallow coast—one great roaring line of breaking waves behind another, till lost in spume, spindrift, and the fierce squalls that were feeding the seas. Mist from their flying tops cut off by the wind filled the great hollows be-*

---

* The current Admiralty chart (3597)—corrected in 2012—warns of "inadequate surveys" around Annenkov Island, and the southern shore of the island itself is marked with a broken line that means it has not been surveyed properly. The chart does, however, reveal an alarming isolated reef that "breaks" about three miles to the west of the island with the cheerful name of "Horror Rock."

> *tween the swells. The ocean everywhere covered by a gauzy tracery of foam with lines of yeasty froth, save where boiling white masses of breaking seas had left their mark on an acre of the surface.*
>
> *On each sea the boat swept upward till she heeled before the droning fury of the hurricane, then fell staggering into the hollow, almost becalmed. Each sea, as it swept us closer in, galloped madly, with increasing fury, for the opposing cliffs, glaciers and rocky points. It seemed but a few moments till it was thundering on the coast beneath the icy uplands, great snow-clad peaks and cloud-piercing crags.*

It was, said Worsley, "the most awe-inspiring and dangerous position any of us had ever been in." It looked as though they were "doomed—past the skill of any man to save." Yet, "with infinite difficulty," they somehow managed to set a small amount of sail and began to claw their way slowly offshore, "praying to heaven that the mast would stand it." The seams of the boat opened as she crashed through the massive seas, and they pumped ceaselessly to keep her afloat. As they looked at the "hellish, rock-bound coast, with its roaring breakers," they wondered—almost dispassionately—at which spot their end was to come. The *James Caird* almost failed to weather the mountainous western extremity of Annenkov Island, which she passed at such close quarters that her crew had to crane their necks to see the snowy peak above them. Once they were beyond it, the wind at last began to moderate. They did well not to attempt the inside passage between Annenkov Island and South Georgia itself as it is encumbered by a dangerous reef that extends for about five miles from its eastern end.

For nine hours, Shackleton and his men had struggled through a hurricane in which, they later learned, a five-hundred-ton steamer on its way to South Georgia was lost with all hands. "I doubt if any of us had ever experienced a fiercer blow than that from noon to 9 P.M.," commented Worsley. By now desperate with thirst and completely exhausted, they finally reached the shore at the entrance to King Haakon Sound. There was no shortage of fresh water but still little opportunity to rest.[13]

Having moved the battered *James Caird* to the head of the sound, Shackleton, Worsley, and one other crew member (Tom Crean, second officer of the *Endurance*) now had to trek well over twenty miles as the crow flies across the high mountains and glaciers that separated them from the settlements on the northern side of the island. The interior of South Georgia consists, as Worsley put it, of a "sheet of ice and snow some hundreds of feet thick except where rocky cliffs and peaks break through."[14] The highest mountains rise over 7,000 feet and one reaches 9,200. The island had only once before been traversed and this was at its narrowest point, where the distance was less than five miles.

Starting out at 3 A.M. on Friday, May 19, 1916, under clear skies and in bright moonlight, they trudged upward and in the light of day they were confronted by a "prospect of spacious grandeur, solitude, and the exquisite purity of Alpine scenery . . . brilliant sunshine on the snow valleys and uplands, with black, upthrusting crags, and peak beyond peak . . . snow-clad and majestic, glittering like armed monarchs in the morning sun."[15] The climb continued, and when the sun dipped behind the mountains the cold was fierce. They finally reached a "razor-back of ice" as darkness closed in. They straddled the ridge, "legs dangling," and debated what to do next while fog rolled in behind them. Having descended for two hundred yards on the far side of the ridge, they stopped and peered into the darkness, unable to see whether the "slope steepened to a precipice or eased out on to the level that seemed so dim and far below." Their chances of survival if the weather worsened would be slight, so Shackleton decided to take the risk of descending by the fastest means:

> *Each coiling our share of the rope beneath us for chafing gear, I straddled behind Sir Ernest, holding his shoulder. Crean did the same to me, and so, locked together, we let go. I was never more scared in my life than for the first 30 seconds. The speed was terrific. I think we all gasped at that hair-raising shoot into darkness. . . . Then, to our joy, the slope curved out, and we shot into a bank of soft snow.*[16]

Pausing occasionally to eat some "hoosh" heated on their Primus stove, they struggled on, finally staggering into the whaling station at Stromness after an almost nonstop journey of thirty-six hours—"a terrible-looking trio of scarecrows." Two young Norwegian lads who bumped into them ran off thinking they were "the devil," but after a bath, shave, and change of clothes, "feeling clean, proud and happy," they had a "royal dinner" with their host, the Norwegian station manager. Worsley at once set off in a whaling vessel to rescue the three men left behind under the *James Caird* in King Haakon Sound, and fell asleep as a gale blew up. "Had we been crossing [the island] that night," he commented, "nothing could have saved us."[17] They afterward learned that there was no other day during the rest of that winter fine enough for them to have traversed the mountains in safety. When Worsley returned with the *James Caird*, the tough Norwegian whalers "would not let us put a hand to her, and every man on the place claimed the honor of helping to haul her up to the wharf." An old captain who "knew this stormy Southern Ocean intimately" told them that it was an honor to shake their hands and declared simply: "These are men!"

Shackleton now faced the challenge of rescuing the twenty-two men left behind on Elephant Island, where they were trying to survive the Antarctic winter under the shelter of two upturned boats on a diet of seal blubber. "During the next hundred days," commented Worsley, "he, Crean and I fought against the elements and every kind of difficulty to effect this purpose." They made four attempts in different vessels. In one small whaler they reached to within sixty miles of their destination but were forced back "by pack-ice, snowstorms and shortage of coal." They then tried in a trawler lent by the Uruguayan government, but were again thwarted by "the accursed pack." From Punta Arenas in southern Chile they next set out in an auxiliary schooner, seventy foot on the waterline:

> *I have commanded small sailing craft in some of the stormiest seas in the world, but that little schooner—with her 40-ft main boom, trying to take charge—flogging her way south from Cape Horn to the pack-ice in the dead of winter, beat them all.*[18]

Again they encountered the ice, the auxiliary engine broke down, and again they were forced to retreat. At last the Chilean navy came to their rescue, and in the little steamer *Yelcho* they groped their way through fog and ice and reached the camp on Elephant Island. As Worsley carefully maneuvered the steamer through icebergs and reefs, Shackleton scanned the beach through binoculars:

> *I heard his strained tones as he counted the figures that were crawling out from under the upturned boat. "Two—five—seven—" and then an exultant shout, "They're all there, Skipper. They are all safe!" His face lit up and years seem to fall off his age.*[19]

Not one man had been lost.[20]

WORSLEY'S EXTRAORDINARY NAVIGATIONAL feat in the *James Caird* depended as much on good judgment as on his skill with a sextant. He knew that his DR calculations were wildly unreliable, and experience told him that he could not safely rely on his latest longitude estimate as they closed the northwestern tip of South Georgia.

Worsley exploited every clue offered by close observation of the natural world around him, though in this he was not unusual. Successful navigation, in the pre-electronic age, depended on the skillful integration of information from many different sources. It was never just a matter of compass, log, and sextant observations: the journals of all the great explorers are full of references to the color of the water, its depth, the nature of the "ground," wave and swell patterns, the clouds, and much else besides. Animal behavior—especially that of birds—was also crucial. FitzRoy commented that he was "not at all surprised that the early voyagers should have taken so much notice of the appearance and flight of birds, when out of sight of land; since in my very short experience I have profited much by observing them, and I am thence led to conclude that land, especially small islands or reefs, has often been discovered in consequence of watching particular kinds of birds, and noticing the direction in which they fly, of an evening, about sunset."[21] Seasoned navigators all over the world instinctively attend to such natural phenomena.

For thousands of years mariners relied on their unaided senses to find their way when they ventured on the open sea, but the native navigators of the Pacific islands were probably the most sophisticated and daring exponents of this kind of "natural navigation." As Bougainville noted with amazement, Polynesian seafarers were able to make successful landfalls without instruments or charts—even on low-lying atolls—after crossing hundreds or thousands of miles of ocean. Modern research[22] has shed a good deal of light on their methods and some of their long ocean passages have been replicated. In 1976, for example, a 65-foot double canoe named *Hōkūle'a*—the Star of Gladness in Hawaiian, or Arcturus—whose design was partly based on drawings of the old voyaging canoes left by Captain Cook, sailed safely from Maui to Tahiti in thirty-one days. She made the journey of 2,500 nautical miles, piloted by a Micronesian navigator named Piailug, who used no instruments of any kind.[23]

From a Western perspective, the most puzzling aspect of the Pacific islanders' navigational methodology is their working assumption that the vessel in which they are sailing is at rest, while the sea and islands "flow" past them. Thomas Gladwin, an expert on the navigators of the Caroline Islands, describes this system as being rather like riding in a train watching the world pass by, only in this case the passing scenery consists of islands:

> *You may travel for days on the canoe but the stars will not go away or change their positions aside from their nightly trajectories from horizon to horizon. . . . Back along the wake, however, the island you left falls farther and farther behind, while the one towards which you are heading is hopefully drawing closer. You can see neither of them, but you know this is happening. You know too that there are islands on either side of you. . . . Everything passes by the little canoe—everything, except the stars by night and the sun by day.*[24]

After long and rigorous training, the native navigators carried in their heads a vast store of knowledge about the rising and setting of the sun and stars, the seasonal behavior of the winds, the relative positions of the different islands, atolls and reefs, and the effects of these on the deep ocean swells as well as on the local, wind-driven waves. They knew exactly where on the horizon each prominent star rose and set and used this information to maintain a constant course when sailing out of sight of land. They may well also have known which star would stand vertically above each important island ("in its zenith") when it crossed that island's meridian. This would have enabled them to determine when they were in the same latitude as the target island, though not whether they were to the east or west of it.[25] The clear nights of the tropical Pacific, which often allow an unimpeded view of the night sky, would—as Bougainville noted—have been a great help to them.

The traditional navigators of Oceania made up for their inability to measure longitude by taking advantage of the distinctive patterns of waves and swells, which revealed to them the presence and the direction of land long before it was visible. When the horizon was obscured and its changing slant could not tell them how their boat was responding to the waves, they apparently stood with their legs apart, using the inertia of their testicles as a guide.[26] Of course they also carefully observed the behavior of birds, the nature of the clouds, and changes in the color of the water. In fact every sense was put to work: sometimes even the taste of the sea could help them fix their position. It was extraordinary skills like these that enabled people not only to settle nearly all the islands of the Pacific, but to develop and maintain a cohesive culture embracing this vast area of ocean over many centuries.[27]

Cook was so impressed by the navigational knowledge displayed by one Tahitian navigator that he agreed to take him aboard the *Endeavour*. Tupia, as he was called, helped Cook explore the neighboring islands and later to communicate with the native Maori population of New Zealand—with whom, it turned out to everyone's amazement, he shared a common language. Sadly he was among those who succumbed to illness contracted in Batavia. Cook and his colleagues do not appear to have made any serious attempt to understand Polynesian navigation, and it was only much later—when the traditional techniques had almost died out—that Westerners began to study the subject seriously.

THE INVENTION OF the sextant allowed the navigator for the first time to attach a numerical value both precise and accurate to the height of a heavenly body above the horizon. It thereby opened up new realms of navigational possibility for Western seafarers. It was a brilliant product of technical ingenuity, but its use still depended on the observer's own eyes, and to obtain the best results skill and practice were required. With the appearance of the chronometer, by contrast, the navigator came to rely on a device that called for nothing from him beyond the ability to wind the mechanism without breaking it. It was a "black box" from which the crucial information issued as if by magic. As the electronic revolution gathered steam during the twentieth century, the navigator's personal contribution to the navigational process diminished still further. With the emergence of GPS this technological "distancing" has reached the point at which fixing one's position is simply a matter of pushing a button.

But there are signs of a reaction setting in. Not everyone wants to be completely in thrall to GPS. The old Polynesian skills are being kept alive by devotees who still practice them today.[28] In fact natural navigation—both by land and by sea—is attracting increasing interest in the West and is once again being taught and practiced.[29] Long voyages without instruments have been made by daring yachtsmen. Perhaps the most remarkable of these was undertaken by Martin Creamer, who set off in 1982 to sail around the world without so much as a compass to steer by, returning home safely after eighteen months.[30] Yet there is no need to go to such extremes to rediscover the challenges and rewards of pre-electronic navigation. The sextant is no more than an extension of the navigator's senses; though far more accurate, it differs not at all in kind from the mariner's astrolabe employed by Mendaña. It offers, I would suggest, a happy compromise: accuracy and reliability coupled with deep immersion in the natural world.

Chapter 18

Two Landfalls

Day 22: Clearer skies and making 5–6 knots in S force 3. At 0600 surrounded by large school of dolphins. A wonderful sight as they kept station around us for about an hour, leaping, diving and criss-crossing under our bows. Beautiful sunshine. A lovely day's sail and more dolphins visited us at 1600. Set spinnaker, which put our speed up to 5 knots. Noon position 49°10' N, 9°20' W. 109 miles run.

We are now back in "soundings"—the depth sounder showed we were in a mere 40 fathoms this afternoon and the water color has changed. The deep blue of the ocean has given way to something greener and murkier and the waves are choppier.

Talked as we all sat in the sun about colonialism—had British done more harm than good? Captain Cook. Tahiti. Harrison's chronometer and lunar distances.

A lovely day and everyone v cheerful.

Day 23: Another grey dawn but the southerly wind still favors us. More ships are passing us as we approach the Scillies. Discussed why we had been set 20 miles to south of DR. Odd—maybe some current. Saw gulls and gannets again for first time in weeks. And lots of trawlers.

Slowly passed south of Scillies at about 20 miles off following Round Island radio beacon on RDF. Picked up Lizard radio beacon and set spinnaker again at 1645.

As night fell we made our landfall—Wolf Rock light (single flash, 30 seconds) looming up on 010°. I turned in and slept, but first we lowered spinnaker to avoid nocturnal disturbances.

Colin and I had taken several sextant sights, as well as bearings of the radio beacons that were now in range, so we knew where we were as we neared the reef-bound Scilly Isles—unlike Shovell. But there are strong tides and many isolated reefs around the Scillies, so accurate pilotage is essential. As the sun went slowly down we scanned the horizon to the northeast for a glimpse of the lighthouse that was our intended landfall. Shortly after sunset it winked into view just where we expected it to be and we all cheered. The next twelve hours, however, were anxious as the almost empty ocean gave way to an overcrowded nautical thoroughfare, full of yachts, fishing boats, and ships heading in almost every direction. For the first time since leaving Halifax, the risk of collision meant that a continuous, careful lookout was vital.

Next morning I came on watch at four in a heavy drizzle, poor visibility, and very light airs. Powerful foghorns were audible all around, and when they came too close I would let off a blast on our own. Once or twice I heard big diesel engines drumming nearer and nearer, but happily they faded safely into the distance. There was very little I could do to avoid passing ships, not least because it was so hard to work out where they were in the murky dawn light, so I just had to hope we were visible on their radar screens.

The sky was lightening and I knew the sun had risen, when quite suddenly—according to my journal—"the mist cleared and I saw the Mulvin Rocks and Lizard Point." The scene of countless shipwrecks, this bold, granite headland was our first sight of land since leaving Halifax. In June 1769 its longitude was fixed exactly for the first time when a party of astronomers sent by Maskelyne observed the Transit of Venus there.[1] Now not only did we know where we were, but we could also see the vessels around us—and there were a lot of them. There was very little wind and we still had plenty of fuel, so we motored most of the way into Falmouth and picked up a visitor's mooring off the town at 11.40 A.M. *Saecwen* showed few signs of her twenty-four-day passage, except for the rust streaks running down her sides, but her crew looked weather-beaten and grubby. Although we were just one of many yachts in an English harbor, we had a secret—we had crossed an ocean.

To step (unsteadily) on dry land, take a long, hot shower and enjoy an uninterrupted night's sleep, to speak again to family and friends—this was all a delight. My mother's excitement at hearing my voice again was touchingly obvious, and from her I learned that the *Tanamo* had indeed reported our position to Lloyd's, who had in turn passed the information on to Colin's family. I vividly recall walking down the narrow high street of Falmouth with Alexa, dodging the crowds and the traffic. We joined a queue to buy ice creams: they tasted wonderful. We felt a deep sense of satisfaction, coupled with relief that the long period of enforced confinement was over and that we had arrived safely, but we also knew that the intense experience we had been through together was a rare one and that quite possibly we would never have such an adventure again.

My transatlantic passage with Colin and Alexa was both a challenge and an education. Not only did I learn how to use a sextant but I also gained new perspectives on time and space, and on my own limitations. Years later I discovered that John Ruskin, though not much of a sailor, was a firm believer in the virtues of the kind of slow, attentive travel in which I had been engaged aboard *Saecwen*:

> *All travelling becomes dull in exact proportion to its rapidity. Going by railroad I do not consider as travelling at all;*

> *it is merely "being sent" to a place, and very little different from becoming a parcel. . . . But if, advancing . . . slowly, after some days we approach any more interesting scenery . . . the continual increase of hope . . . affords one of the most exquisite enjoyments possible to the healthy mind; besides that real knowledge is acquired of whatever it is the object of travelling to learn, and a certain sublimity given to all places, so attained, by the true sense of the spaces of earth that separate them.*[2]

Ruskin's remarks apply with even greater force to modern modes of travel, and our stately progress in *Saecwen*—though seaborne—illustrated his point to perfection. Of course I was intellectually aware of the size of the ocean before we set out from Halifax, but spending twenty-four days crossing it under sail gave its dimensions a very different and truly sublime reality. The long night watches looking up at the stars in the black immensity of space were a lesson in humility, and the experience of a gale in mid-Atlantic left me wondering what it must be like to encounter a proper storm. People often talk idiotically about "conquering" mountains or "defying" the sea, but there is no contest. I was left with an overwhelming sense of nature's vast scale and complete indifference, and this had a strangely calming effect. We come and we go, the earth too was born and will eventually die, but the universe in all its chilly splendour abides.

Alcyone, April 1981

In 1980 I became the part owner of a Contessa 32 built by Jeremy Rogers. These legendary boats are widely acknowledged to be among the best modern cruising yachts—a classic design with beautiful lines, but strong and seaworthy, even if they offer less ample accommodation than many boats of a similar size. Tough people have sailed Contessa 32s into high Arctic and Antarctic waters, and around Cape Horn. We called our new boat *Alcyone*, after the brightest star in the Pleiades, and equipped her with long-distance cruising in mind.

Alcyone was launched in Lymington on March 1, 1980, and we spent the first season getting to know our new boat. There were many weekend excursions to local haunts like Chichester Harbor, Newtown on the Isle of Wight, Poole and Lulworth Cove, followed by a longer trip to the Channel Islands, Normandy, and Brittany. My wife's brother, Chris, and I had ambitiously put our names down for the Two-Handed Transatlantic Race of 1981, but that proved unrealistic: not only would the passage take up to a month, but the preparations would also absorb several weeks and we would then have to bring the boat back. It was going to be impossible to take so much time off, so we had to reconsider. We decided to aim for the Azores, a widely scattered group of volcanic islands in the middle of the Atlantic in roughly the same latitude as Lisbon.

Colin had often spoken of the beauty of these islands, and at a distance of 1,300 nautical miles they were far enough away for the voyage to be a genuine challenge. During the early part of 1981 we began making preparations—which included the purchase of a new Freiberger sextant and some exotic Portuguese charts—and then in March I had commitments in New York. Returning in mid-April it was a rush to get everything ready in time for our planned departure. On the sixteenth Chris and I drove down to Lymington and loaded the boat. At 11:30 on a clear, cold night the two of us set sail for Cherbourg, where we planned to stock up with duty-free liquor and other vital provisions. We streamed the Walker log as we passed the Needles lighthouse at 12:30 A.M. and headed south across an almost empty channel in a freshening northeasterly breeze. It was very cold. I soon started to feel seasick and regretted the fierce curry we had eaten earlier, but it was a quick passage and we reached Cherbourg by ten the next morning.

Having done all our shopping the day before, on the morning of Saturday, April 18, we filled up with water and made our farewell telephone calls. This was before the days of satellite phones and *Alcyone* carried only a short-range VHF radiotelephone, so from now on we would be out of touch until we reached Ponta Delgada, in perhaps two weeks' time. Though *Alcyone* was a good, strong boat and the distance to the Azores was only half the width of the Atlantic, I felt just as I had on the eve of our departure from Nova Scotia back in 1973. I was very conscious that this time I was going to be responsible for the navigation—and that I had not been able to practice with the new sextant.

The wind was moaning through the rigging and the halyards of the yachts tied up around us in the marina were slatting noisily against their masts. It was blowing a brisk force 7 from the northeast and the air was icy, but it was a perfect wind for our purposes and we were not going to waste it. So at 11:50 A.M. we motored out into the great harbor behind the massive nineteenth-century breakwaters, hoisted the storm jib, passed out through the western entrance, and shot off down the channel with both wind and tide hurrying us on. By 3 P.M. the island of Alderney was abeam, and at nightfall we were well on our way. At 10 P.M. we were steering 240° and the wind had moderated to force 4, though it was still very cold. Chris and I took turns wearing a large woollen balaclava that was very welcome on our night watches. The log records simply: "Orion on starboard bow and Alcyone!" The wind eased during the night and at dawn Chris raised the big genoa in place of the much smaller jib. When I came on deck later that morning the sun was shining, and we had a large breakfast of scrambled eggs, French bread, and coffee.

Soon after 1 P.M. on the second day out, when the sun was almost due south of us, I brought out the sextant and tried my hand at a mer alt. It gave our latitude as 49°02', which seemed plausible. That was encouraging, but later in the day when I tried my first timed sight to generate a fix, the complexities of the sight-reduction tables defeated me. That was worrying, but I now also discovered that in our hurried departure we had forgotten to make a note of the frequencies of the powerful radio beacons in the Azores before we set sail. Designed for transatlantic aircraft, and with a range of several hundred miles, these would have made our navigational task simple. Now it was too late to obtain them and we would have to rely exclusively on celestial navigation to find our way. We were back in the nineteenth century.

This was a much bigger challenge than I had expected. What if we missed the islands completely and ended up sailing to South America? What if we ran out of food and water? All sorts of crazy fears ran through my head. For an hour or so as we headed steadily westward I was ready to turn back, but, encouraged by Chris and a shot of whisky, I got out my navigation books, did some homework, and found the solution to my problem with the tables. It was actually quite simple after all. During the night we used RDF to fix our position and the wind moved into the east, now a comfortable force 4–5. A full moon broke through the clouds and at 5:45 A.M. I took the altitude of Polaris, which gave our latitude as 48°25.9'. Then at 9:30 A.M. on day three I took a sun sight and drew my first position line on the plotting chart. Yes, I could still do it! Now I could relax. At noon the mer alt gave us our first celestial fix: 48°09' North, and 7°57' West.

We were crossing the continental shelf and would soon be in deep, oceanic water. Our battery-driven quartz watch (not clockwork this time) was one second slow by the radio time signal at 2 P.M. The weather was sunny with a few clouds, and gannets were folding their wings as they dive-bombed the sea around us hunting their prey. Chris made goulash for supper, and then we shared the night watches—four hours on, four off. The northeasterly wind was still cold and the balaclava remained in use. At 2:20 A.M. a ship passed us going in the opposite direction, and I tried to raise her, but without success. A Polaris altitude at 5 A.M. gave a latitude of 47°35' North, and I noted in the log that the barometer was slowly rising, which was a good sign.

As we headed south and west, the weather grew milder and the sea changed from a sludgy gray-green to clear, dark indigo. This was blue-water sailing. The old daily rhythm soon returned—taking a morning sight, followed by the mer alt and then perhaps one or two further sights in the afternoon or at dusk. The log records sights of the moon, Jupiter, Alkaid, and Arcturus, as well as Polaris and, of course, the sun. The helpful northerly winds persisted and we made fast progress, the little circles that marked our daily position steadily advancing across the chart. On the sixth day out we passed through an enormous raft of by-the-wind-sailors. A few inches across and with a little bluish sail on top, these small medusas—closely related to the Portuguese man-of-war—carpeted the surface of the sea so thickly that they washed aboard every time we dipped our bows into a wave. The imprisoned IRA activist Bobby Sands monopolized the BBC news, which we were still picking up. Terrible though it was, his hunger strike seemed far more remote than the seven hundred miles we had travelled could explain.

After eight days at sea we were nearing our goal, well ahead of schedule. The winds had eased, the air was warm, and we were reaching gently through calm seas. As darkness fell, the skies were cloudless and the waning moon rose in the early hours amid a blaze of stars and planets—the brilliant Jupiter and Saturn almost in conjunction ahead of us in the southwest, Regulus in the west, Alkaid and Arcturus high above us, Antares low in the southeast, Vega, Deneb, and Altair off in the north and east. The horizon was well defined, and I sat up half the night taking sights, partly for the fun of it: it was a good sextant and my skills were returning. Lots of position lines appeared on the plotting chart and it looked as if we were being set to the east by a current of about half a knot. We were now steering for the western end of São Miguel, the main island of the Azores. The water was very clear, and, as we whispered along in the soft breeze, I stood in the cockpit watching as our wake streamed out far behind us—a long, green-glowing milky furrow, sparkling in the darkness with flashes of planktonic fire.

Suddenly I saw three or four luminous, pale green torpedo tracks heading fast toward us. Just as they were about to hit, they shot downward, disappearing under the boat. I looked over the side in astonishment but for half a minute or so could see nothing. Then deep, deep down a faint group of lights began to corkscrew vertically upward, growing brighter and brighter until at last they broke the surface close alongside in a shower of sparks. They were dolphins, and now I could hear their voices. I called Chris up on deck and we both watched—entranced—for half an hour or more as they created aquatic arabesques beneath and around us, as if for our delight. It was at once the most beautiful and the most mysterious spectacle: a flight of angels could hardly have been a greater surprise, or felt more like a blessing, but as suddenly as they had arrived, the dolphins departed, leaving us bereft. We kept watch, hoping they might return, but they did not. Perhaps our company was too dull.

A landfall, according to Conrad, may be either good or bad:

> *In all the devious tracings the course of a sailing-ship leaves on the white paper of a chart she is always aiming for that one little spot—maybe a small island in the ocean, a single headland upon the long coast of a continent, a lighthouse*

on a bluff, or simply the peaked form of a mountain like an ant-heap afloat upon the waters. But if you have sighted it on the expected bearing, then the Landfall is good.[3]

The following morning, Monday, April 27, after consulting the log and taking a sun sight, I raised my head from the chart table and announced that high land should soon appear fine on the port bow, at a distance of about twenty-five miles. Within an hour, at 9:50 A.M., the cliffs and hills of São Miguel obligingly emerged from the blue haze, and as we drew near we caught the sweet scent of the trees and damp soil that covered them. "Land on the nose!" Chris wrote in the log, and it was, in truth, a good landfall. Although I had known our position, and knew that I knew, I was delighted and relieved to see the physical proof. How many sailors before me have felt as I did? The light of the sun and stars—reflected in the twin mirrors of a sextant—had shown us the way, as it had countless others before us.

In warm sunshine we rounded the lighthouse of Santa Clara, on the west end of São Miguel, at 3:30 P.M. and followed the south coast until the massive breakwater that protects the harbor of Ponta Delgada came into view. At 4:30 we anchored in fourteen feet of water, just nine days out from Cherbourg and having covered nearly 1,300 nautical miles by the log at an average speed of 5½ knots. The waterfront looked exotic with its palm trees and white stucco houses. Once ashore, we went through the tedious bureaucratic processes of clearing customs and border control, and then found our way through the old town to the São Pedro Hotel—a grand old building in the Portuguese style, its long, elegant windows and doors framed in grey volcanic rock.

Chris and I were tired, dirty, and happy. We had arrived several days earlier than planned and had—even if unintentionally—navigated in the old-fashioned way, relying entirely on compass, log, clock, and sextant. We made phone calls to announce our safe arrival, got cleaned up, and ate a big dinner in the elegant, paneled dining room, looking out over the harbor, where *Alcyone* rode at anchor in the light evening breeze. A vase filled with fresh bird of paradise flowers reminded us that we had left a cold English spring far behind. Perhaps we should have drunk a toast to all those who had made our modest achievement possible, but we were too pleased with ourselves to think of it.

It now seems right to repair that omission. So, belatedly, I offer my tribute to the generations of astronomers, mathematicians, and instrument makers who brought celestial navigation to perfection, and to the dedicated and courageous mariners who charted the world's oceans with a sextant in their hands. To salute them in the words with which Matthew Flinders remembered his beloved Trim might seem eccentric, but I think they would understand:

Peace be to their shades, and Honour to their memory.

Epilogue

As *Saecwen* was sailing across the Atlantic in August 1973, the United States Air Force was on the point of developing the satellite navigation system that would before long supplant the sextant: GPS. The first phase of the program was approved in December 1973, though of course Colin and I had no idea that the skills he was teaching me would soon be outmoded.[1]

Although GPS was not the first satellite navigation system, it was far more sophisticated than its predecessors, and the challenges facing its developers were formidable. The engineers had to design and build more than twenty-four satellites, each carrying an atomic clock specially shielded against the effects of cosmic radiation, all of which then had to be delivered by rocket into carefully calibrated orbits. Each of these satellites transmits extremely precise information about its location in space, together with a time signature (accurate to a few nanoseconds). By comparing the signals received from four or more of them, a GPS receiver anywhere on the surface of the earth can fix its position to within a few yards, as well as determining its velocity and elevation. So sensitive is the system that it has to allow for the distorting effects on the onboard clocks of the high velocities of the satellites—in accordance with the predictions of Einstein's special theory of relativity—as well as the pressure exerted on each satellite by the light of the sun.[2] Although the service started to function in a limited form in the early 1980s, it was not until the mid-1990s that GPS became fully operational. It is reported to have cost the U.S. government $12 billion to develop, and its maintenance is also expensive; in a remarkably generous spirit, the U.S. government made GPS freely available to the public, worldwide, in 1983, though—for reasons of national security—only the U.S. military had access to the most accurate positional data until May 2000, when the system of "selective availability" was ended. The Russians have developed a similar satellite system—GLONASS—and other nations are following their example, as is the European Union.

Though the technology it employs is vastly different, GPS bears a conceptual resemblance to the "new" celestial navigation described in Chapter 15. A GPS fix is, in effect, the product of the analysis of equal-altitude circles[3] derived from the signals each satellite transmits. The array of satellites is designed to ensure that at least four satellites are always "visible" above the user's horizon at any time, anywhere in the world. Each satellite can be likened to a star that emits not light, but radio signals that reveal its precise position in the sky. Unlike the light from a star, however, these signals are also stamped with the exact time of their transmission. While the celestial navigator deploys almanac and sight-reduction tables to define and then solve the trigonometric problems arising from a sextant "sight," the GPS user relies on computer algorithms in the receiver to perform analogous calculations—instantly, effortlessly, and automatically.

I first encountered satellite navigation when I was sailing across the South China Sea from Hong Kong to Manila in April 1984. I was navigating with the sextant I had used going to and from the Azores three years earlier, though the experience of working out sights under a tropical sun that passed almost vertically overhead at noon was very different. Beyond the strong winds of the northeast monsoon we were becalmed off the coast of Luzon. I have never been so hot. Shoals of flying fish burst from the sea, sometimes crash-landing on our deck. During the night, ceaseless lightning flashes turned the tall thunderclouds over the land into colossal, flickering Chinese lanterns, but they were so distant that no sound of thunder reached us as we gently rolled on the still sea. Next day the heat and humidity were so intense that when I tried to draw a position line on the chart the damp paper ripped, and it was a relief to be able to turn on the satellite navigation set. It was an earlier, less sophisticated system than GPS known as TRANSIT, and it took some time before enough satellites came "up" over our horizon to generate a fix, but after a few trials it became clear that the new technology was indeed extraordinarily accurate.

Paradoxically, the very accuracy of satellite-based navigation can give rise to navigational problems. Positions marked on paper charts often fail to match those delivered by GPS, and sometimes the discrepancies are surprisingly large. I have anchored in a Hebridean bay only to discover that, according to the satellite fix, the boat should have been halfway up a neighboring mountain. It is easy to see that clashes of this kind could lead to disaster. The trouble is not just that the old surveyors, relying on their sextants, were unable to fix their positions with the precision of GPS, or that mapping the surface of a sphere onto a flat sheet of paper results in distortions: these are indeed serious issues, but the problems run deeper. The earth is not a perfect sphere; as discussed earlier, it is flattened toward the poles, but it also exhibits other, more complex irregularities that must be taken into account if positions on its surface are to be accurately charted. These detailed eccentricities were not appreciated when the first charts were made and anyway would not have signified much when navigational methods yielded fixes accurate (at the very best[4]) to within a few hundred yards or so.

Unfortunately, however, there are several different models of the earth's shape, each of which may yield a slightly different position for a given set of coordinates. In the age of GPS, it is essential to know which of these models your chart is based on and to adjust your receiver's parameters accordingly.

The old paper charts—which are bulky and expensive—are gradually disappearing as navigators rely increasingly on "electronic chart display and information systems" (ECDIS) that can be loaded on computers and readily updated. Connected directly to the GPS, these "moving map displays" show the vessel as a little ship moving across the surface of the "chart," and provide very precise information about its course and speed, as well as the effects of currents and leeway, the state of the tide, not to mention distances to, and estimated times of arrival at, any chosen point. Nothing could be more convenient, but such "electronic charts" reduce the navigator's role almost to that of a spectator—or perhaps, more precisely, a mere consumer of navigational data.

In the light of these developments, there has been much debate in nautical circles about the continuing relevance and usefulness of celestial navigation. The U.S. Navy has taken the bold step of ceasing to train all officers in the use of the sextant: only navigational specialists now have that privilege. In most national navies and the merchant service, however, courses in celestial navigation are still required. Safety is usually advanced as the reason for retaining the old skills. At the simplest level, it is obvious that electronic systems—like GPS—can and often do break down, for example when electricity supplies fail or equipment is accidentally damaged. But there are other risks, too.[5] Charged particles emitted by the sun can give rise to disturbances of the earth's magnetosphere capable of disrupting electronic equipment in orbit—or even on the ground. Although we usually have some warning of such events, and the GPS network is designed to be resilient, there is always the possibility that some satellites might be accidentally disabled. The system is also vulnerable to acts of war, and those who control it can easily restrict access to it.

Perhaps a more pressing danger, however, arises from inexpensive jamming devices designed to block the reception of the very weak signals emitted by GPS satellites. Drivers of commercial vehicles fitted with GPS trackers sometimes use such jammers to conceal from their employers exactly where they have been, but in so doing they may unintentionally block GPS coverage over a wide radius. Nor is the vulnerability of GPS of concern only to navigators: many of the electronic systems on which modern life depends—including cellular phones and the computers that manage international financial markets—rely on GPS time signals. In view of all these real or potential problems, an "enhanced" modern version of LORAN is being developed to provide a robust and accurate backup to GPS.[6]

STRANGELY ENOUGH, THE work of an artist—James Turrell—has helped me understand why navigating with a sextant casts such a powerful spell over me. Turrell has spent half a lifetime turning an extinct volcano in a remote corner of the Painted Desert of Arizona into what can justly be described as a celestial work of art. Roden Crater is a naked-eye observatory, without telescopes or computers, where the only sounds are the wind in the juniper bushes and the crunch of the reddish pumice under foot. Turrell's medium is light itself, and the "skyspaces" he has built within the crater are designed to help us catch sight of ourselves in the very act of seeing and to discover with freshly opened eyes the everyday wonders of light—especially the light that comes from beyond the "ocean of air" that is the earth's atmosphere. To enter one of these skyspaces is revelatory. When I was lucky enough to meet him, I was not surprised to discover that Turrell is an aviator and a sailor, that he knows how to find his way by the light of the heavens. His art is embedded in his reverence for the natural world.

Turrell describes the light that reaches us from outer space as "old," in contrast to the "new" light that we generate ourselves. It is a nice distinction, though in cosmic terms the light from the stars we can see with our naked eyes is *extremely* young. Our Milky Way—filled with hundreds of billions of stars—is just one of countless galaxies, many of which are so distant that their light, when at last it reaches us, has been travelling since long before the earth coalesced out of a cloud of cosmic dust some 4.5 billion years ago, and the stars from which it comes may long since have died. In 2012 astronomers reported observing a galaxy formed not long after the Big Bang itself that is (or was when its light set out on its journey) more than 13 billion light-years distant from us.[7] Thirteen billion years travelling through space at roughly 186,000 miles a second: that is a long, long way. Yet it was only when the first big reflecting telescopes were built early in the last century that anyone knew for sure that the Milky Way was not a solitary archipelago of light in a pitch-black, otherwise empty universe. The huge new mirrors resolved the mysterious blurry nebulae revealed by the old telescopes into vast congregations of stars. The known universe had just grown vastly bigger, and it has gone on growing as the reach of telescopes has continued to increase.

Our species is thought to have first emerged about two hundred thousand years ago. Almost everything about our prehistory remains uncertain, but there is no doubt that until very recently our ancestors survived in the wild as hunter-gatherers, living off the land or sea. The details must have varied a good deal from place to place, as they do among the few groups that still survive in these ways, but in every case people were immersed in the natural world, and profoundly influenced by it, if not completely under its control. Survival depended on thoroughly understanding the environment in which they lived—when and where the food and water they needed could be found, the timing of migrations of birds and animals, the behavior of predators, seasonal weather patterns, and so on. With the development of agriculture and settled life in villages and towns over the last ten thousand years, humans have achieved a growing measure of independence from the vagaries of nature. But until the Industrial Revolution almost everyone was engaged directly or indirectly in the production of the raw materials of life: food, clothing, and shelter. Very few lived in towns, most of which were still so small that the countryside would never have felt remote.

Over the last few hundred years the balance has gradually but decisively changed. Today most of the world's population live in cities, many of which have grown so large that their inhabitants have little or no day-to-day contact with the natural world. They cannot even see the stars at night. The accelerating growth of urban culture has been closely associated with the development of new technologies, many of which have contributed to our growing alienation from "wild nature": new methods of transport and communication, new kinds of food and drink, new forms of work. In short, we now have completely new ways of living. A frequently observed feature of modern life is that it makes fewer and fewer physical demands, with the result that more and more of us are suffering from illnesses associated with a sedentary existence. We are also being gradually relieved of the mental challenges that were once part of everyday life. Computers eliminate the burden of performing the simplest calculations, they correct our spelling and grammar, help us to remember what we have to do, "manage" the engines of our cars, help us take photographs, and much more besides. Now they even do our navigation for us.

If asked, most sailors would probably agree that they endure the many discomforts of life at sea in order to find spiritual refreshment in the largest wilderness on the planet. They might also mention the attractions of self-sufficiency and independence. It is then ironic that so many navigators should now choose to find their way at sea by watching a little red ship moving over a bright blue screen rather than by using the natural cues around them. In doing so, they are not only turning their backs on the very things that make the whole undertaking worthwhile, but they are also denying themselves the precious rewards of agency—the use of hand, head, and eye to solve problems and overcome difficulties.[8]

GPS and all the other electronic gadgetry certainly offer reassurance and convenience, but the solutions they provide are delivered without the slightest effort on our part, and often with as little understanding. And excessive reliance on such technology may actually weaken our natural navigational skills.[9] As GPS threatens to return us all to the helpless dependency of childhood, offshore sailors in particular run the risk of neglecting an older, more demanding, yet far more personally rewarding technology. I am not calling for electronic navigational tools to be abandoned, but we should avoid becoming exclusively dependent on them. It is time to rediscover the joys of celestial navigation, not merely as a safety net, but because using a sextant to find our way puts us in the closest possible touch with the natural world at its most sublime.

Outside the abstract domain of mathematics, nothing better symbolizes the timeless order and perfection of the divine realm than the workings of the cosmos. When I recall learning how to handle a sextant all those years ago, I see myself, a transient speck of life, fixing my position on the surface of our small planet by taking the measure of vast, unimaginably distant suns whose lives are measured in billions of years. The chastening contrast between their calm majesty and my fretful pettiness was overwhelming. And yet, small as I was, when I caught the stars in the mirrors of my sextant, I felt as if I was entering into a strange communion with them. After crossing countless miles of space, light emitted by those colossal thermonuclear fires had at last reached *my* eyes, had told me exactly where *I* was. When I look up at the stars in the night sky that once showed me the way across an ocean, a sense of wonder engulfs me, and I bow my head.

Acknowledgments

For reasons that will be obvious, I should first record my thanks to my father and to Colin McMullen. I am also deeply grateful to my mother for encouraging me to embark on that life-changing voyage aboard *Saecwen* so many years ago.

Special thanks are due to my agent, Catherine Clarke, for helping me to refine the original proposal for this book and for persuading HarperCollins to publish it. Arabella Pike, my editor, saw the point of what I was trying to do and has been very supportive throughout the book's development while at the same time offering shrewd and constructive criticism. I am immensely grateful to her, and to the team who have put the book together—Jo Walker, who designed the cover; Katherine Josselyn and Tara Al Azzawi who have masterminded the PR and marketing; Peter James, the copy editor; Geraldine Beare who assembled the index; and especially Kate Tolley, who has overseen the whole process with such skill, patience, and good humour.

I have benefited from conversations and email exchanges with Dr. Richard Dunn, Senior Curator and Head of Science and Technology at the National Maritime Museum, who has also kindly read and commented on parts of the manuscript. Richard's colleagues on the Cambridge Digital Library Board's Longitude Project have helped me with technical queries: Katy Barrett, Dr. Rebekah Higgitt, and Dr. Nicky Reeves. I am also grateful to the staff of the Caird Archive and Library at the National Maritime Museum, the National Archives, the United Kingdom Hydrographic Office, and the British Library.

The Director of the Royal Institute of Navigation, Peter Chapman-Andrews, has been very helpful, and I have made extensive use of the Institute's Cundall Library as well as the Institute's online archives. Peter put me in touch with David Rydiard, staff author, *Admiralty Manual of Navigation*, who kindly read and commented on the technical parts of the manuscript. J. D. Hill of the British Museum gave me useful advice about the Nebra Sky Disc. I am also grateful to Professor Maya Jasanoff of Harvard University and Professor Claudio Aporta of Dalhousie University for helpfully responding to my queries.

I would also like to thank Tristan Gooley and John Heilbron for their advice and encouragement, and Javier Mendez Alvarez of the Roque de Los Muchachos Observatory on La Palma in the Canary Islands, who showed me how a modern observatory works and patiently answered many questions. Warner Bros. kindly gave permission for me to quote from the screenplay of *Mutiny on the Bounty*.

Heather Howard, Colin McMullen's daughter, and Vanessa de Mowbray have generously allowed me to reproduce photographs of Colin and of *Saecwen*, while my sister Fiona Rogers, my nephew Kit Rogers, and his wife, Jessie Lane, have commented helpfully on the manuscript. My sister-in-law, Elizabeth Gibson, has given me valuable advice, and her husband, Rick Morgan, brought to my attention and translated the passage from the *Lusiads* quoted in Chapter 3. My wife's cousin, Jane Kimber, and her husband, Jonny Clothier, have been extremely generous in allowing me to borrow their beautiful yacht, *Brown Bear*, in various parts of the world.

My wife, Mary, and my daughters, Eleanor and Miranda, have cheerfully endured my enthusiasm for celestial navigation over many years and have encouraged me greatly during the writing of *Sextant*. But it is to Mary, above all, that I owe my deepest thanks—not least for her expert help in polishing the manuscript.

Notes

INTRODUCTION

1 Beaglehole (1966) 41ff.
2 Ibid. 36.
3 Ibid. 57.
4 According to Mendaña's own estimate, the "Western Islands" lay some 5,000 nautical miles from Peru, but this was more than 2,000 miles short. Such large errors were by no means unusual: one of Magellan's pilots was out in his estimate of the longitude of the Philippines by almost 53 degrees—equivalent to some 3,000 nautical miles. See Beaglehole (1966) 38.
5 Beaglehole (1966) 80.
6 Ibid. 317ff.
7 Jean-Nicolas Buache was later to be custodian of the French hydrographic service—the Dépôt des Cartes et Plans de la Marine: Van der Merwe 45.
8 The sextant was a development of Hadley's "quadrant," which continued in use for many years after the sextant was introduced. But the name sextant has long been used as a generic term for all instruments of this kind, and that is the sense in which I use it here.
9 Lack of space prevented the inclusion of other remarkable hydrographers—like Alessandro Malaspina, who led the great Spanish expedition to the Pacific of 1789–94, or William Fitzwilliam Owen, who surveyed much of the coast of Africa in the 1820s while also conducting campaigns against slave traders.
10 See, e.g., Lewis.
11 GPS is in fact just one satellite navigation system, though by far the best known and most widely used. The generic term for all such systems is Global Navigation Satellite Systems, or GNSS.

CHAPTER 1: SETTING SAIL

1 Beaglehole (1974) 33–34.
2 Ritchie 15.
3 Beaglehole (1974) 54.

4 Ibid. 58.
5 Ibid. 71–72.
6 Conrad 2.

CHAPTER 2: FIRST SIGHT

1 Hakluyt VI.38–42.
2 Howse and Sanderson 109.
3 Translations given in Bowditch 248.
4 Mary Blewitt, *Celestial Navigation for Yachtsmen* (5th ed., 1973).
5 Homer, *The Odyssey*, Book 5, lines 269–75: my translation.
6 After making the necessary corrections to the reading (for sextant error, "height of eye," atmospheric refraction, the sun's semi-diameter and parallax), the sun's altitude was subtracted from 90 degrees to give the "zenith distance." To this figure was added the value of the sun's (northerly) declination to yield the latitude. If the sun's declination had been southerly—as it is between late September and late March—it would have been subtracted from the zenith distance. The sun's declination should also be corrected for the approximate longitude of the ship, as even a difference of a few hours from Greenwich can make a small but significant difference.

CHAPTER 3: THE ORIGINS OF THE SEXTANT

1 Joshua Slocum, *Sailing Alone Around the World* (1956).
2 James L. Gould, "Animal Navigation: A Galaxy of Cues," *Current Biology*, vol. 23, 4 (2013) 149–50; Marie Dacke, Emily Baird, Marcus Byrne, Clarke H. Scholtz, and Eric J. Warrant, "Dung Beetles Use the Milky Way for Orientation," ibid. 298–300.
3 P. Kraft, C. Evangelista, M. Dacke, T. Labhart, and M. V. Srinivasan, "Honeybee Navigation: Following Routes Using Polarized-Light Cues," *Philosophical Transactions of the Royal Society B*, vol. 366, 1565 (2011) 703–708.
4 Jonathan T. Hagstrum, "Atmospheric Propagation Modeling Indicates Homing Pigeons Use Loft-Specific Infrasonic 'Map' Cues," *Journal of Experimental Biology*, vol. 216 (2013) 687–99.
5 B. Mauck, N. Gläser, W. Schlosser, and G. Dehnhardt, "Harbour Seals (Phoca vitulina) Can Steer by the Stars," *Animal Cognition*, vol.11, 4 (2008) 715–18.
6 Aveni 87–92.
7 Taylor 99–100.
8 Cotter (1968) 35.
9 Ibid. 56.
10 Luís Vaz de Camões, *The Lusiads*, Canto V, verses 25–27, translation courtesy of Rick Morgan.
11 Cotter (1968) 67.
12 Newton's bitter rival Robert Hooke had invented a simpler instrument, on similar lines and employing only a single mirror, as early as the 1660s, whereas Newton's relied on the use of two mirrors: see Cotter (1968) 74–75.

13 Ibid. 77ff.
14 Ibid. 81.
15 Ibid.
16 Bruyns 106, for example.
17 Melville 448–49.

CHAPTER 4: BLIGH'S BOAT JOURNEY

1 Beaglehole (1974) 498.
2 Bligh 160.
3 Bligh claimed in his published account of the voyage that he was allowed to take only a "quadrant," but his journal shows that he actually had a sextant and navigational books.
4 Bligh 156–64.
5 Ibid. 177.
6 Ibid. 180.
7 Ibid. 182.
8 Ibid. 192.
9 Ibid. 197.
10 Ibid. 220.
11 Ibid. 227.
12 Ibid. 234.
13 Ritchie 190.

CHAPTER 5: ANSON'S ORDEALS

1 This famous sea battle took place in May 1941. The *Bismarck* was the largest European battleship of the day, and the pride of the German navy. She was attempting to break out into the North Atlantic with her consort, the heavy cruiser *Prinz Eugen*, when she was intercepted in the Denmark Strait by the elderly British battle cruiser *Hood* and the brand-new battleship *Prince of Wales*. *Hood* exploded and sank in a matter of minutes after being straddled by a German broadside that probably ignited one of her magazines. More than 1,400 men died—only a handful survived. A shell from *Prince of Wales*, however, punctured one of *Bismarck*'s fuel tanks, forcing her to cut short her mission and head for France. She was sunk two days later with heavy loss of life. *Prince of Wales* herself was later sunk by Japanese bombers off the Malay Peninsula. Colin, who was among the last to leave the ships, swam from the bridge as she turned over. He dryly commented that they were lucky not to have been attacked by sharks while they floated in the sea waiting to be rescued.
2 The Battle of Jutland took place on May 31–June 1, 1916. It was the only full-scale encounter between the British and German fleets of World War I. The risks were enormous, especially for the British. Winston Churchill had chillingly warned the British commander-in-chief, Sir John Jellicoe, that he was the one man on either side who could lose the war in a single afternoon. If the British Grand Fleet were to suffer serious losses, it would no longer be possible to maintain the cru-

cial naval blockade of Germany, and if Britain itself were then cut off, the war would very soon come to an end—with a German victory. It was essential that the British fleet should survive largely intact, however desirable it might be to destroy the German one. It was therefore vitally important that Jellicoe adopt the right tactics, and this depended on knowing not only where the enemy ships were, but also where his own forces stood in relation to them. When the Grand Fleet sailed from its base in Orkney, it was responding to intelligence—based on intercepted radio signals—that German warships were leaving harbor and heading out into the North Sea. Seventeen hours later, as the two vast armadas converged off the coast of Denmark, DR errors had accumulated, and poor visibility had—according to Colin—prevented navigation officers aboard the various different British units from fixing their positions by sextant sights. At the same time those of Jellicoe's subordinates who were already in contact with the German fleet failed to pass on any useful information about its course, speed, and position. The commander-in-chief was therefore struggling to interpret conflicting navigational information about the whereabouts of his own forces, while remaining uncertain where the enemy fleet—now approaching him at a combined speed of 50 miles an hour or more—was going to appear. As much by luck as by good judgment, Jellicoe maneuvered the Grand Fleet into exactly the right position to intercept the German fleet. The battle itself—though bloody—was not strategically decisive, but if Jellicoe had made the wrong decision about how to deploy his battleships, it might have been a catastrophe for the British. It is sobering to reflect that one of the factors affecting the outcome of the battle—and therefore conceivably of the entire war—may have been the misty North Sea weather, which precluded the use of the sextant. Celestial navigation remained as important in 1916 as in the days of Cook, and it was to remain so until the 1940s.

3 Nowadays most yachts crossing an ocean would at least be equipped with an Emergency Position-Indicating Radio Beacon (EPIRB)—a device that automatically sends out a distress signal that can be picked up by search-and-rescue satellites. They might also carry a satellite telephone. We had none of these.

4 An ingenious mechanism, governed by a wind vane and powered by a blade beneath the water, that controls the tiller by means of a system of ropes and pulleys.

5 May 27ff.

6 See Dash.

7 Dash 5–6.

8 Many of the so-called soldiers were in fact retired veterans of earlier wars, most of whom were elderly and unfit for any kind of service. See Williams (1999) 19–22.

9 Anson 22.

10 It was not until the end of the eighteenth century that the crucial importance of fresh fruit and vegetables in preventing scurvy was finally recognized by the Royal Navy. See Williams (1999) 224–26.

11 Anson 144–45.

12 Ibid. 115.

13 Ibid. 117.
14 Williams (1999) 47.
15 Ibid. 47–49.
16 Anson 117–18.
17 Ibid. 149.
18 Ibid. 150.
19 Ibid. Anson's instructions placed the island in the wrong latitude and much too close to the coast. See Williams (1999) 54. Williams states that the *Centurion* first tried to reach the island by "sailing down its latitude," but this is inconsistent with the account quoted here, and with the track chart in the authorized account of the voyage.
20 Anson 151.
21 Ibid. 152. See also May 29ff. A young officer named John Campbell was among Anson's crew. In 1757 he was to be given the task of testing the practicality of the lunar-distance method of determining longitude at sea.
22 Anson 393.
23 Williams (1999) 139.
24 Ibid. 216–18.

CHAPTER 6: THE MARINE CHRONOMETER

1 Although they were certainly accomplished seafarers, we know little about the wayfinding techniques the Greeks and Romans employed and it has been assumed that they had no special navigational instruments, relying instead on their skill and experience—rather like the Polynesian navigators. However, recent research suggests that the hitherto mysterious Antikythera device, recovered in 1900 by sponge divers from a shipwreck off southern Greece that dates from the first century BCE, employed a remarkably complex system of gears to reveal how the sun, moon, and five planets known at that time behaved. If the Greeks were capable of producing mechanisms like this, we may have underestimated their navigational technology.
2 May 9.
3 Heilbron 155–56, 167–70.
4 The first of these—*La Connoissance* [*sic*] *des Temps*—was published in France in 1678.
5 Andrewes 94.
6 Other research conducted on the same expedition revealed that the strength and direction of the earth's gravitational field varied from place to place. This result had important implications for surveyors. It also meant that the rates of pendulum clocks would be subject to slight but significant variations.
7 The Loch of Stenness, on Mainland. Curiously enough, one end of this baseline was anchored by the prehistoric stone circle of the Ring of Brodgar.
8 Quoted in Cotter (1983).
9 Danson 180.
10 Ibid. 15.

11 Cotter (1968) 192ff.; Howse 11–13.
12 Heilbron 235–36.
13 Ibid. 346–48.
14 Ibid. 348.
15 Williams (1994) 93.
16 Part of the problem with the first trial was that the longitude of Port Royal (the harbor in Jamaica where the tests were carried out) was still uncertain.
17 Howse 125.
18 Andrewes 282ff.

CHAPTER 7: CELESTIAL TIMEKEEPING

1 Cotter (1968) 199ff.
2 Ibid. 202.
3 Ibid. 81.
4 Hewson 85.
5 James Bradley, letter to Mr. Cleveland, Secretary of the Admiralty, April 14, 1760, reproduced in Mayer cxi–cxv.
6 Howse 127–28.
7 Ibid. 27–39.
8 Cotter (1968) 206.
9 Ibid.
10 Howse 157, 200.
11 Ibid. 50–52.
12 Ibid. 221.
13 Ibid. 51.
14 Ibid. 83.
15 Quill 171–73.
16 Howse 77–79, 125–26.
17 These followed the example set by the French astronomer Nicolas-Louis de Lacaille, who had made lunar observations at sea in the early 1750s. Lacaille's own, rather less accurate, lunar tables were first published in the *Connoissance des Temps* for 1761. See Howse 41.
18 Rodger 623.
19 Howse 94.
20 Ibid. 93–94.
21 Norie 226.
22 Raper xii.
23 Ibid. 127.

CHAPTER 8: CAPTAIN COOK CHARTS THE PACIFIC

1 Whitfield 105.
2 Howse and Sanderson 93, 125.
3 Beaglehole (1974) 102ff.
4 Fry 121.

5 Beaglehole (1974) 87–90.
6 Ibid. 145.
7 Ibid. 134.
8 Ibid. 137.
9 Ben Finney, "James Cook and the European Discovery of Polynesia," in Fisher and Johnston 20.
10 John Elliott quoted in Beaglehole (1974) 361–62.
11 Beaglehole (1955–69) II.304–305.
12 Forster 441.
13 Beaglehole (1955-69) II.322.
14 Ibid. 643.
15 Forster 446–47.
16 Beaglehole (1974) 335.
17 Beaglehole (1955–69) I.343–44.
18 Beaglehole (1962) II.81.
19 Beaglehole (1955–69) I.353.
20 Ibid. 375.
21 Ibid. 377–78.
22 Ibid. 378.
23 Beaglehole (1962) II.106.
24 Beaglehole (1955–69) I.379.
25 Ibid.
26 Beaglehole (1974) 246.
27 Ibid.
28 The National Archives, ADM 51/4545.
29 Beaglehole (1974) 154.
30 Beaglehole (1955–69) I.9n.
31 Ibid. II.61n.
32 Hough 205.
33 Beaglehole (1955–69) II.cxii.
34 Ibid. 525.
35 Ibid.
36 Ibid.
37 Forster 239.
38 Beaglehole (1955–69) II.665.
39 Ibid. 692.
40 Ibid. cxi n.
41 Beaglehole (1974) 129.
42 Rodger 382–83.

CHAPTER 9: BOUGAINVILLE IN THE SOUTH SEAS

1 Bideaux and Faessel 5.
2 Bougainville (1771) 16.
3 Cook was sniffy about the lack of navigational detail in the first edition of the

Voyage, but this deficiency was largely remedied in the expanded, second edition of 1772.

4 See Bideaux and Faessel 19ff.
5 Bougainville (1772) I.214.
6 Bougainville (1771) 120.
7 Suthren 108, 132.
8 Bougainville (1771) 155.
9 Ibid. 158–61.
10 Ibid. 170.
11 Ibid. 185.
12 Ibid. 189–90.
13 Ibid. 190.
14 Ibid. 191.
15 Ibid. 195–96.
16 Ibid. 253–54.
17 Bideaux and Faessel 260n.
18 Bougainville (1771) 194.
19 Ibid. 198.
20 Ibid. 209.
21 Ibid. 197.
22 Ibid. 199, 201.
23 Ibid. 219.
24 Ibid. 216.
25 Ibid. 227–28 (1771).
26 Ibid. 200ff.
27 He was not, however, using the lunar-distance method. See Bideaux and Faessel 221n.
28 Bougainville (1771) 209.
29 Ibid. 255.
30 Ibid. 256.
31 Ibid. 257.
32 Ibid. 258.
33 Ibid. 261.
34 Ibid. 263.
35 Ibid. 278–79.
36 Bougainville (1772) I.xxxii.

CHAPTER 10: LA PÉROUSE VANISHES

1 Dunmore 104–105.
2 Ibid. 127.
3 Ibid. 177–78.
4 Milet-Mureau I.246–48.
5 Ibid. 8.
6 Ibid. 43.

7 Ibid. 159.
8 Ibid. II.46.
9 Ibid. I.252.
10 Ibid. II.7.
11 Dunmore (1994) I.lxx.
12 Milet-Mureau I.13–61.
13 Ibid. II.161ff.
14 Ibid. 178.
15 Ibid. 301.
16 Ibid. 302.
17 Ibid. 303. Similar fruitless searches for reported reefs or rocks in midocean—known as *vigías* from the Spanish word for "look out"—were to absorb the energies of explorers for many years to come.
18 Ibid. 306.
19 Uncle of the builder of the Suez Canal.
20 Milet-Mureau III.154.
21 Ibid. 141.
22 Ibid. 186ff.
23 Ibid. 188.
24 Ibid. 189–90.
25 Ibid. 191.
26 Ibid.
27 Ibid. 198.
28 Ibid. 200–205.
29 Letter to Lecoulteux de La Noraye, February 7, 1788, quoted in Dunmore (1985).
30 Dunmore (1994) I.xxx.
31 Ibid. ccxl.
32 Dillon 1.33–34. The island was then also known as Vannicolo.
33 Ibid. 34–35.
34 Ibid. II.166–67.
35 Ibid. 194–95; 216–18.
36 Ibid. 306.
37 Ibid. 205.
38 Ibid. 254–55.
39 Ibid. 398–99.
40 Ibid. 400–403.
41 Ibid. 397.
42 Dunmore (1985) 295–96.
43 Dunmore (1994) I.lviii.

CHAPTER 11: THE TRAVAILS OF GEORGE VANCOUVER

1 Coleman 13.
2 Ibid. 21.
3 Ibid. 32.

4 Ibid.
5 Ibid. 35.
6 Ibid. 38.
7 Vancouver I.v–vi.
8 Bougainville (1771) 17.
9 Vancouver I.xxix.
10 Raban 50–51.
11 Coleman 41.
12 Ibid. 40.
13 Ibid. 47, 50.
14 Ibid. 50–51.
15 Vancouver I.195–6–vi.
16 Quoted by Andrew David, "Vancouver's Survey Methods and Surveys," in Fisher and Johnston 51–52.
17 David, "Vancouver's Survey Methods and Surveys," in ibid. 52–53.
18 See Nicholas A. Doe, "Captain Vancouver's Longitudes—1792," *Journal of Navigation*, vol. 48 (1995) 374–88. Doe has demonstrated that Vancouver could have discovered these errors when he returned home by comparing the predictions in the *Nautical Almanac* with the actual observations made by the astronomers at Greenwich on the dates in question, as Flinders was later to do on his return from Australia.
19 Andrew David discusses Vancouver's navigational difficulties in "Vancouver's Survey Methods and Surveys," in Fisher and Johnston 64–67.
20 Vancouver I.319–20.
21 Ibid. 321.
22 Ibid. 363–64.
23 Ibid. 420ff.
24 Coleman 98, 135.
25 Fisher and Johnston 7–8.
26 Coleman 139–43.
27 Coleman 149–50.

CHAPTER 12: FLINDERS—COASTING AUSTRALIA

1 Scott 33–37.
2 Ingleton 10.
3 Baker 7. Quotation from letter to Sir Joseph Banks, December 8, 1806.
4 Scott 12–13.
5 Ibid. 19.
6 Ibid. 29.
7 Ibid. 44ff.
8 It is often claimed that Flinders invented the name "Australia," but the word was apparently first used in the late seventeenth century in a translation from a French work of fiction. See Baker 107.
9 Flinders I.xcvi.

10 Ibid. xcvii.
11 Ibid. xcix–xcx.
12 Ibid. cxiii.
13 Ibid. cxvii.
14 Scott 108.
15 Flinders I.cxix–cxx.
16 Ibid. cxxxviii.
17 Named after Alexander Dalrymple.
18 Ibid. clxiii.
19 Ibid. clxxi.
20 Ibid. clxxxii ff.
21 Ibid. cxciii.
22 Scott 154–55.
23 Flinders I.3–5.
24 Ibid. 4.
25 National Maritime Museum, Flinders Papers FLI/25/4.
26 Ibid. FLI/25/5.
27 Ibid. FLI/25/2.
28 Ibid. FLI/25/6.
29 Ingleton 99.
30 National Maritime Museum, Flinders Papers FLI/25/7.
31 Flinders I.7.
32 Ingleton 110.
33 National Maritime Museum, Flinders Papers FLI/25.
34 Flinders I.9.
35 Ibid. 10
36 Letter to Ann Flinders, 31 May 1802: National Maritime Museum, Flinders Papers FLI/25/14.
37 Flinders I.106.
38 Ibid. 189.
39 Ibid. 190.
40 National Maritime Museum, Flinders Papers FLI/25/14.
41 Flinders I.230.
42 Ibid. 229.
43 Flinders II.48–49.
44 Ibid. 26.
45 Ibid. 143.
46 National Maritime Museum, Flinders Papers FLI/25/18.

CHAPTER 13: FLINDERS—SHIPWRECK AND CAPTIVITY

1 Flinders II.299.
2 Ibid. 309n.
3 Ibid. 312–13.
4 Ibid. 315.

5 Ibid. 321.
6 Ibid. 327.
7 Ibid. 323.
8 Ibid. 351.
9 Ingleton 267.
10 Flinders II.360–64.
11 Ingleton 340.
12 Brown and Dooley 120.
13 Ingleton 340–41.
14 National Maritime Museum, Flinders Papers FLI/25/26.
15 Flinders II.485.
16 Flinders I.iv–v.
17 Ibid. 256.
18 Ibid. 258.
19 Ibid. iii.
20 Ibid. 255.
21 Ibid. iii.
22 Ibid. 259.
23 Ibid. 261.
24 Ingleton 419.
25 Ibid. 420.
26 Ibid. 421.
27 Ibid. 423.
28 Flinders II.333.
29 National Maritime Museum, Flinders Papers FLI/11.
30 Sterne 95.
31 National Maritime Museum, Flinders Papers FLI/11.

CHAPTER 14: VOYAGES OF THE *BEAGLE*

1 "On First Looking Into Chapman's Homer."
2 FitzRoy I.xix.
3 Ibid. 73.
4 Ibid. 72.
5 Ibid. 74.
6 Ibid. 77.
7 Ibid. 78.
8 Ibid. 79.
9 Ibid. 179.
10 Ibid. 180–81.
11 Ibid. 153.
12 Ibid. 218.
13 Ibid. 222.
14 They were then encamped at the entrance to Skyring Water. See FitzRoy I.229.
15 Ibid. 231–32.

16 Ibid. 240–41.
17 Ibid. 203.
18 Ibid. 363.
19 Ibid. 373.
20 Ibid. 383.
21 Ibid. 432.
22 Ibid. 434.
23 Ibid. 435–36.
24 Gribbin and Gribbin 119, 129, 136.
25 FitzRoy II.25–26.
26 Ibid. 344.
27 Ibid.
28 Ibid. 371–72.
29 Gribbin and Gribbin 256.
30 Ibid. 273.
31 Ibid. 274.
32 Ibid. 290.

CHAPTER 15: SLOCUM CIRCLES THE WORLD

1 It may be a little unfair to credit Sumner exclusively with this "discovery," as several earlier navigators had made similar observations: see Cotter (1968) 271–75.
2 Cotter (1968) 277.
3 Ibid. 278.
4 R. Ch. Duval, "Admiral Marcq de Blond de St. Hilaire," *Journal of Navigation*, vol. 19 (1966) 209–12.
5 Cotter (1968) 293ff.
6 The exact steps are as follows: First draw a line through the assumed position along the bearing of the computed azimuth. Then compare the observed altitude with the computed altitude; if the observed altitude is greater than the computed one it follows that your actual position is closer to the GP of the observed body than the assumed one—and vice versa. Then convert the difference in altitude into a distance in nautical miles (this is called the "intercept"—one minute of arc equaling one nautical mile), and mark this on the azimuth line, either toward or away from the GP. At this point draw a line at right angles to the azimuth line. This represents a tiny segment of one of the equal-altitude circles discussed earlier, and you are somewhere on it: it is a "line of position."
7 C. H. Cotter, "A Centennial Tribute to Marcq St. Hilaire," *Journal of Navigation*, vol. 28 (1975) 449ff.
8 Real professionals (like hydrographers) need also to know the biases that they bring to the task of taking a sight. Most of us cannot, even with long practice and in ideal circumstances, measure a sextant angle (or record the exact time of a sight) with complete accuracy, and our errors tend to follow a pattern. Once

these biases are known, they can be allowed for by applying a "personal equation" to the sights we take.

9 Quoted in Cotter (1968) 306

10 The merchant service was much slower to adopt it, however, long remaining faithful to the old "longitude by chronometer" method.

11 Lecky 462.

12 Wolff 79–80.

13 Ibid. 93–94.

14 Ibid. 96–97.

15 Ibid. 97–98.

16 Ibid. 100–111.

17 Slocum 22.

18 Ibid. 15.

19 Ibid. 24.

20 Ibid. 96.

21 Ibid. 98–99.

22 Ibid. 100.

23 Ibid. 101–102.

24 *South America Pilot* II.167.

25 Slocum 102.

26 FitzRoy III.307.

27 Slocum 102.

28 Ibid. 142.

29 Ibid. 144–45.

30 Ibid. 146–47.

31 Ibid. 148.

32 Ibid. 149.

33 Bowditch iv, 233.

34 Wolff 206–209.

CHAPTER 16: *ENDURANCE*

1 Thomson 22.

2 Thomson 37.

3 Shackleton 7.

4 Ibid. 32–33.

5 Ibid. 39.

6 Ibid. 52.

7 Ibid. 57.

8 Ibid. 63.

9 Ibid. 67.

10 Ibid. 71–72.

11 Ibid. 72.

12 Ibid. 79–80.

13 Ibid. 81.

14 Ibid. 82.

15 Ibid. 83.
16 Worsley 24.
17 Ibid. 25.
18 Ibid. 27.
19 Ibid. 30.
20 Ibid. 32–33.
21 Ibid. 41.
22 Ibid. 44–45.
23 Ibid. 46–47.
24 Ibid. 50.
25 Ibid. 54–55.

CHAPTER 17: "THESE ARE MEN"

1 Worsley 60–61.
2 Ibid. 62.
3 Shackleton 174.
4 Worsley 64.
5 Shackleton 181.
6 Worsley 67–68.
7 Ibid. 82.
8 Ibid. 74–75.
9 Ibid. 86.
10 Ibid. 88.
11 Ibid. 92.
12 Ibid. 93–94.
13 Ibid. 95–99.
14 Ibid. 120.
15 Ibid. 125–26.
16 Ibid. 128–29.
17 Ibid. 139–41.
18 Ibid. 145.
19 Ibid. 144–46.
20 However, the members of Shackleton's expedition who were sent ahead to establish a depot for him on the other side of the Antarctic did not fare so well. Two men died and four others only narrowly survived: see Shackleton 265–66.
21 FitzRoy I.557–58.
22 See Lewis 1994. David Lewis was an intrepid single-handed sailor, and his account of the navigational skills of the Pacific islanders is full of fascinating details.
23 Lewis 313ff.
24 Quoted in ibid. 176.
25 Ibid. 280ff.
26 Ibid. 127.
27 Comparable skills are exercised by the Inuit in the northern Canadian Arctic. Relying on close observation of a complex array of natural cues, including

stars, winds, snowdrifts, tides and animal behavior, they can find their way across wide expanses of sea, ice, and land with complete confidence—in almost all weather conditions. But it takes years of "quiet tutoring and experience" to gain these skills, and there are concerns that increasing reliance on GPS will erode them. See Claudio Aporta and Eric Higgs, "Satellite Culture: Global Positioning Systems, Inuit Wayfinding, and the Need for a New Account of Technology" in *Current Anthropology*, vol. 46, issue 5, December 2005, 729–53.

28 The Polynesian Voyaging Society, for example. See their website: http://hokulea.org/vision-mission.

29 See, for example, Gooley; Huth.

30 See Gooley. 158ff.

CHAPTER 18: TWO LANDFALLS

1 Howse 111–12.

2 Ruskin V.370–71.

3 Conrad 2.

EPILOGUE

1 See Edward M. Lassiter and Bradford Parkinson, "The Operational Status of NAVSTAR/GPS," *Journal of Navigation*, vol. 30, issue 01, January 1977, 3–47.

2 Solar pressure alone can displace a satellite by almost 40 meters each day.

3 Strictly speaking, spheres, since GPS operates in three dimensions.

4 Modern estimates of the accuracy of celestial fixes obtained at sea suggest that an error radius of about two nautical miles is to be expected—if the observer is experienced, the conditions are good, the time is precisely known, and the sextant is reliable. (See, e.g., N.L.A. Bovens, "Position Accuracy of Celestial Fixes," *Journal of Navigation*, vol. 47, issue 02, May 1994, 214–20.) Multiple observations at a fixed point on land can, however, reduce this to a few hundred yards.

5 For a much fuller discussion of the vulnerabilities of GPS and other satellite navigation systems, see Dr. Wolfgang Schuster, "Protecting the Future," *Navigation News, The Magazine of the Royal Institute of Navigation* (September–October 2013) 22–24.

6 "Ships' navigators go back to the future as white van man gets them into a jam," *Times*, March 30, 2013.

7 *New Scientist*, December 12, 2012.

8 Matthew Crawford has written very interestingly about this subject—though in a slightly different context. See Crawford esp. 59–61.

9 For an interesting overview of this subject, see Alex Hutchinson's article "Global Impositioning Systems—Is GPS technology actually harming our sense of direction?" in *Walrus*, November 2009 (http://thewalrus.ca/global-impositioning-systems/). See also Nicholas Carr, "All Can Be Lost: The Risk of Putting Our Knowledge in the Hands of Machines," in *Atlantic*, November 2013 (http://www.theatlantic.com/magazine/archive/2013/11/the-great-forgetting/309516/).

Glossary

Azimuth: the bearing of a celestial body's geographical position measured in degrees either from the north or south pole (whichever is nearer the observer).

Beam: the width of a ship at her widest part; also a term for the transverse members on which a ship's deck is laid.

Beam ends: A ship is on her beam ends when she is heeled over by ninety degrees.

Binnacle: the housing for the steering compass, which usually includes a small lamp.

Bosun's chair: a board on which someone can be hoisted aloft to carry out repairs or inspect the rigging.

Broach-to: A vessel running before a strong wind broaches to when the helmsman loses control and she turns sharply to windward. She may then be "knocked down" by the pressure of the wind in her sails.

Cable: a distance of two hundred yards.

Careen: heaving a ship over onto her side to expose her bottom for purposes of repair or maintenance.

Chains (or "cheans"): a small platform on either side of a ship from which the lead-line is heaved.

Cockpit: a well in the deck of boat—usually close to the stern and uncovered—from which she is steered, either by tiller or by wheel.

Declination: the angle of a heavenly body north or south of the equator as measured from the center of the earth.

Embayed: A sailing vessel is embayed when blown by onshore winds into a bay from which it cannot escape by *tacking*.

Fathom: six feet.

Genoa: a very large triangular sail set on the main forestay of a yacht.

Geographical position: the point on the earth's surface vertically beneath a heavenly body.

Great circle: a circle on the surface of the earth that has its center at the center of the earth. The equator is one example, and all *meridians* are parts of great circles.

Greenwich Hour Angle: the angle between the Greenwich meridian and the meridian over which a heavenly body is passing at any given moment.

Jib: a generic term for triangular sails set on a forestay of a ship or yacht.

Larboard: the old term for "port" (as opposed to "starboard").

Lead-line: a specially marked rope to which is attached a lead weight for measuring the depth of water and for taking samples of the sea bottom (using a bit of sticky grease or tallow).

Leeward: downwind, as opposed to upwind or "windward." Hence *lee shore*.

Local Hour Angle: the angle between the observer's meridian and the meridian over which a heavenly body is passing at any given moment.

Log: any device for measuring a ship's speed through the water. The traditional "chip log" consisted of a weighted, wooden quadrant that, when thrown overboard, pulled out a carefully measured "log-line" marked—literally—with "knots." By counting the number of "knots" that passed in a given interval of time (usually determined by the "log glass") it was possible to estimate the vessel's speed in "knots" or nautical miles per hour.

LORAN: short for LOng RAnge Navigation, a radio-based navigation system developed in the United States during the World War II for military purposes. By measuring the time intervals between pulsed transmissions from shore-based radio stations, it is possible to fix a ship's position on a special chart. An enhanced form known as E-LORAN is being developed as a backup to GPS.

Main (or *mainsail*): the principal sail of any sailing vessel. In most yachts it is attached to the mainmast and extended by a long horizontal spar known as the "boom."

Meridian: half of a great circle joining the north and south geographical poles, otherwise known as a line of longitude.

Nautical mile: the length of one minute of arc of a great circle at the earth's surface, or approximately 6,080 feet. It is 15 percent longer than a statute mile.

North: true north marks the direction of the geographical north pole; magnetic north marks the direction of the magnetic north pole.

Offing: a safe distance offshore, avoiding all dangers.

Paying off: The bows of a sailing vessel "pay off" when they have passed through the eye of the wind and start to drop off to *leeward*, typically when *tacking*.

Quarter: The after part of a ship is divided into two quadrants on either side of the midline—the port and starboard quarters. Hence *quarter gallery*: a small gallery on each quarter of a ship connecting with the stern cabin often accommodating a lavatory, and *quarterdeck*, the after part of the upper deck of a ship, usually reserved for the captain and officers.

Schooner: typically a two-masted sailing vessel in which the foremast is lower than the mainmast.

Scuttle (or "Skuttel"): a circular port or window in the side of a ship, commonly called a porthole.

Shrouds: the fixed (or "standing") rigging used to support the mast of a ship laterally. The "stays" by contrast provide fore and aft support.

Staysail: a triangular sail set on one or more of the forestays of a ship or yacht.

Steerage: a large cabin below the *quarterdeck* and just forward of the great cabin.

Stream anchor: a spare anchor, often deployed from the stern of a ship.

Tacking: the process of bringing the bows of a sailing vessel up into the wind and then allowing the sails to fill on the other side or "tack" when going to windward.

Tender: a small vessel serving in support of a larger one.

Tiller: a long bar attached to the head of the rudder by means of which a boat is steered. Larger vessels are steered with wheels.

Tonnage: Originally a measure of cargo capacity, a ship's tonnage was estimated in the days of sail by a formula taking into account the length and breadth ("beam") and depth ("draft") of a vessel. This is *not* a measure of the weight of a vessel, which is known as its "displacement."

Whaleboat: a rudderless open boat, propelled by oars and pointed at both ends. Modelled on those used for hunting whales, they were steered with an oar and could easily be beached. They were particularly useful in coastal survey work.

Yard: a spar on which sails are carried; in a square-rigged ship the yards cross the masts horizontally.

Yawl: a two-masted sailing boat in which the smaller, after mast (or "mizen") is mounted (or "stepped") very close to the stern.

Zenith: the point in the sky vertically above the observer.

Zenith distance: the angular distance between a heavenly body and the observer's zenith.

Bibliography

Admiralty Manual of Hydrographic Surveying (1965 ed.). 2 vols. London: Hydrographer of the Navy.

Admiralty Manual of Navigation, Volume 2 (1973 ed.). London: Her Majesty's Stationery Office.

Andrewes, W. J. (ed.) (1998). *The Quest for Longitude* (2nd ed.). Cambridge, MA: Harvard University Press.

Anson, G. (1748). *A Voyage Round the World . . . by George Anson Esq . . . compiled by Richard Walter MA* (2nd ed.). London: John & Paul Knapton.

Aughton, P. (undated). *Voyages That Changed the World*. London: Quercus.

Aveni, A. (2008). *People and the Sky: Our Ancestors and the Cosmos*. London: Thames & Hudson.

Baker, S. J. (1963). *My Own Destroyer: A Biography of Matthew Flinders*. Sydney: Angus & Robertson.

Bauer, B. (1995). *The Sextant Handbook* (2nd ed.). Camden, ME: International Marine.

Beaglehole, J. C. (ed.) (1955–74). 4 vols. *The Journals of Captain James Cook on His Voyages of Discovery*. Cambridge: Hakluyt Society.

Beaglehole, J. C. (ed.) (1962). *The Endeavour Journal of Joseph Banks 1768–1771*. 2 vols. Sydney: Angus & Robertson.

Beaglehole, J. C. (1966). *The Exploration of the Pacific* (3rd ed.). Stanford, CA: Stanford University Press.

Beaglehole, J. C. (1974). *The Life of Captain James Cook*. London: A. & C. Black.

Bideaux, M., and Faessel, S. (eds.) (2001). *Louis-Antoine de Bougainville: Voyage autour du monde: édition critique*. Paris: Presses de l'Université de Paris-Sorbonne.

Blake, J. (2004). *The Sea Chart: The Illustrated History of Nautical Maps and Navigational Charts*. London: Conway Maritime Press.

Blewitt, M. (1973). *Celestial Navigation for Yachtsmen* (5th ed.). London: Edward Stanford.

Bligh, L. W. (1792). *A Voyage to the South Sea*. London: George Nicol.

Bougainville, L. A. (1771). *Voyage autour du monde par le frégate La Boudeuse et la flûte L'Étoile* (1st ed.). Paris: Saillant & Nyon.

Bougainville, L. A. (1772). *Voyage autour du monde par le frégate La Boudeuse et la flûte L'Étoile* (2nd ed.). 2 vols. Paris: Saillant & Nyon.

Bowditch, N. (2002). *The American Practical Navigator*. Bethesda, MD: National Imagery and Mapping Agency.

Brown, A. J., and Dooley, G. (eds.) (2005). *Matthew Flinders' Private Journal*. Adelaide: Friends of the State Library of South Australia.

Bruyns, W. F. (2009). *Sextants at Greenwich*. Oxford: Oxford University Press.

Cobbe, H. (ed.) (1979). *Cook's Voyages and Peoples of the Pacific*. London: British Museum.

Coleman, E. C. (2006). *Captain Vancouver: North-West Navigator*. Stroud: Tempus.

Conrad, J. (1906). *The Mirror of the Sea*. London: Methuen.

Cotter, C. H. (1968). *A History of Nautical Astronomy*. London: Hollis & Carter.

Cotter, C. H. (1983). *A History of the Navigator's Sextant*. Glasgow: Brown, Son & Ferguson.

Crawford, M. (2009). *The Case for Working with Your Hands, or Why Office Work Is Bad for Us and Fixing Things Feels Good*. London: Penguin Books.

Danson, E. (2006). *Weighing the World: The Quest to Measure the Earth*. Oxford: Oxford University Press.

Dash, M. (2003). *Batavia's Graveyard: The True Story of the Mad Heretic Who Led History's Bloodiest Mutiny*. London: Weidenfeld and Nicholson.

Dawson, L. S. (1885). *Memoirs of Hydrography*. Eastbourne: Henry W. Keay.

Dillon, P. (1829). *Narrative and Successful Result of a Voyage in the South Seas . . . to Ascertain the Actual Fate of La Pérouse's Expedition*. 2 vols. London: Hurst, Chance, & Co.

Douglass, D., and Hemingway-Douglass, R. (1996). *Exploring the North Coast of British Columbia*. Bishop, CA: Fine Edge Productions.

Dunmore, J. (1985). *Pacific Explorer: The Life of Jean François de La Pérouse*. Palmerston North, NZ: Dunmore Press.

Dunmore, J. (ed.) (1994–95). *The Journal of Jean-François de Galaup de la Pérouse, 1785–1788*. 2 vols. Trans. J. Dunmore. London: Hakluyt Society.

Fagan, B. (2012). *Beyond the Blue Horizon: How the Earliest Mariners Unlocked the Secrets of the Oceans*. London: Bloomsbury.

Fernández-Armesto, F. (2006). *Pathfinders: A Global History of Exploration*. Oxford: Oxford University Press.

Fisher, R., and Johnston, H. J. (eds.) (1993). *From Maps to Metaphors: The Pacific World of George Vancouver*. Vancouver: University of British Columbia Press.

FitzRoy, R. (ed.) (1839). *Narrative of the Surveying Voyages of His Majesty's Ships the Adventure and Beagle*. 3 vols. London: Henry Colburn.

Flinders, M. (1814). *A Voyage to Terra Australis*. 2 vols. and atlas. London: W. Bulmer.

Fry, H. T. (1970). *Alexander Dalrymple and the Expansion of British Trade*. London: Frank Cass.

Gooley, T. (2010). *The Natural Navigator*. London: Virgin Books.

Gribbin, J. and M. (2003). *FitzRoy: The Remarkable Story of Darwin's Captain and the Invention of the Weather Forecast*. London: Yale University Press.

Hakluyt, R. (1927). *The Principal Navigations, Voyages, Traffiques and Discoveries of the English Nation*. 8 vols. London: J. M. Dent.

Heilbron, J. L. (2010). *Galileo*. Oxford: Oxford University Press.

Hewson, J. B. (1983). *A History of the Practice of Navigation* (2nd ed.). Glasgow: Brown, Son & Ferguson.

Hoare, M. E. (ed.) (1982) *The* Resolution *Journal of Johann Reinhold Forster 1772–1775*. 4 vols. London: Hakluyt Society.

Hough, R. (1995). *Captain James Cook: A Biography*. London: Hodder & Stoughton.

Howse, D. (1989). *Nevil Maskelyne: The Seaman's Astronomer*. Cambridge: Cambridge University Press.

Howse, D., and Sanderson, M. (1973). *The Sea Chart*. Newton Abbot: David & Charles.

Huth, J. E. (2013). *The Lost Art of Finding Our Way*. Cambridge, MA: Harvard University Press.

Ingleton, G. C. (1986). *Matthew Flinders, Navigator and Chartmaker*. Guildford: Genesis in association with Hedley.

Johnson, D. S., and Nurminen, J. (2007). *The History of Seafaring: Navigating the World's Oceans*. London: Conway Maritime Press.

Kemp, P. (ed.) (1979). *Oxford Companion to Ships and the Sea*. Oxford: Oxford University Press.

Lecky, S. T. (1903). *"Wrinkles" in Practical Navigation* (14th ed.). London: George Philip & Son.

Lewis, D. (1994). *We, the Navigators: The Ancient Art of Landfinding in the Pacific* (2nd ed.). Ed. S. D. Oulton. Honolulu: University of Hawaii Press.

Mack, J. (2011). *The Sea: A Cultural History*. London: Reaktion Books.

May, W. E. (1973). *A History of Marine Navigation*. Henley-on-Thames: G. T. Foulis.

Mayer, T. (1770). *Tabulae Motuum solis et lunae*. London: Richardson.

Melville, H. (1974). *Moby-Dick, or The Whale*. London: Folio Society.

Milet-Mureau, M. L. A. (1797). *Voyage de La Pérouse autour du monde*. 4 vols and atlas. Paris: Imprimerie de la République.

Mixter, G. W. (1960). *Primer of Navigation* (4th ed.). Princeton, NJ: Van Nostrand.

Norie, J. W. (1839). *A Complete Epitome of Practical Navigation*. London: J. W. Norie.

Quill, H. (1966). *John Harrison: The Man Who Found Longitude*. London: John Baker.

Raban, J. (2000). *Passage to Juneau: A Sea and Its Meanings*. London: Picador

Raper, L. H. (1840). *The Practice of Navigation and Nautical Astronomy*. London: R. B. Bate.

Ritchie, R. A. (1967). *The Admiralty Chart: British Naval Hydrography in the Nineteenth Century*. New York: Elsevier.

Rodger, N. A. (2005). *The Command of the Ocean: A Naval History of Britain 1649–1815*. London: Penguin Books.

Ruskin, J. (1903–12). *The Works of John Ruskin*. 39 vols. Ed. E. T. Cook and A. Wedderburn. London: George Allen.

Scott, E. (1914). *The Life of Captain Matthew Flinders RN*. Sydney: Angus & Robertson.

Shackleton, S. E. (1999). *South: The Endurance Expedition*. London: Penguin Books.

Skelton, R. A. (1970). *Explorers' Maps: Chapters in the Cartographic Record of Geographical Discovery*. London: Hamlyn.

Slocum, J. (1956). *Sailing Alone Around the World*. New York: Dover Publications.

Sobel, D. (1996). *Longitude: The True Story of a Lone Genius Who Solved the Greatest Scientific Problem of His Time*. London: Fourth Estate.

South America Pilot Volume II (1993 ed.). Taunton, UK: Hydrographic Office.

Sterne, L. (2005). *The Life and Opinions of Tristram Shandy, Gentleman.* London: Folio Society.

Suthren, V. (2004). *The Sea Has No End: The Life of Louis-Antoine de Bougainville.* Toronto: Dundurn.

Taylor, E. G. (1956). *The Haven-Finding Art: A History of Navigation from Odysseus to Captain Cook.* London: Hollis & Carter.

Thomson, J. (1999). *Shackleton's Captain: A Biography of Frank Worsley.* Oakville, Ontario: Mosaic Press.

Vancouver, J. (ed.) (1798). *A Voyage of Discovery to the North Pacific Ocean and Round the World, in the Discovery Sloop of War and Armed Tender Chatham, under the Command of Captain George Vancouver.* 3 vols. London: G. G. and J. Robinson & J. Edwards.

Van der Merwe, P. (ed.) (2003). *Science and the French and British Navies 1700–1850.* London: National Maritime Museum.

Van Dorn, W. G. (1975). *Oceanography and Seamanship.* London: Adlard Coles.

Whitfield, P. (1996). *The Charting of the Oceans.* London: British Library.

Williams, G. (1999). *The Prize of All the Oceans.* London: HarperCollins.

Williams, J. E. (1994). *From Sails to Satellites: The Origin and Development of Navigational Science.* Oxford: Oxford University Press.

Wilson, E. O. (1984). *Biophilia: The Human Bond with Other Species.* Cambridge, MA: Harvard University Press.

Wolff, G. (2011). *The Hard Way Around: The Passages of Joshua Slocum.* New York: Vintage.

Worsley, C. F. (1974). *Shackleton's Boat Journey.* London: Folio Society.

Illustrations

Plate sections:

Saecwen on the coast of Maine (*Courtesy Louise de Mowbray*)

The Nebra Sky Disc (©*Anagoria*)

Star-trails in the northern sky (© *Babak Tafreshi, The World at Night*)

Aquatint of the mutineers turning William Bligh and his crew from the *Bounty* by Robert Dodd, 1790 (© *National Maritime Museum, Greenwich, London*)

Pastel portrait of William Bligh in 1791 by John Russell RA (*Courtesy of the Captain Cook Memorial Museum, Whitby*)

Chalk drawing of Nevil Maskelyne attributed to John Russell RA, *c.* 1776 (© *National Maritime Museum, Greenwich, London*)

Sextant made in London by John Bird *c.* 1758 (© *National Maritime Museum, Greenwich, London*)

A page from Charles Green's journal (© *The National Archives*)

Lithograph of Louis-Antoine de Bougainville after a portrait by Zépherin Belliard, date unknown (*Courtesy National Library of Australia*)

Portrait of Jean-François de Galaup, comte de La Pérouse, by J. B. Greuze (*All rights reserved: Toulouse-Lautrec Museum—Albi—Tarn—France*)

Probable portrait of George Vancouver by an unknown artist, 1796–98. Oil on canvas (© *National Portrait Gallery, London. NPG 503*)

"The Caneing in Conduit Street" by James Gillray. Hand-coloured etching, published by Hannah Humphrey October 1, 1796 (© *National Portrait Gallery, London. NPG D12578*)

Portrait of Captain James Cook by John Webber RA, 1782. Oil on canvas (*Collection: National Portrait Gallery, Canberra. Purchased 2000 by the*

Integrated images:

Index

Note: Page numbers in *italics* indicate photographs and illustrations.

S
M
I
M'
S'
80
70
60